A BRIEFER HISTORY OF TIME

A BRIEFER HIS

TORY OF TIME

STEPHEN HAWKING

WITH LEONARD MLODINOW

BANTAM PRESS

LONDON · TORONTO · SYDNEY · AUCKLAND · JOHANNESBURG

TRANSWORLD PUBLISHERS
61–63 Uxbridge Road, London W5 5SA
a division of The Random House Group Ltd

RANDOM HOUSE AUSTRALIA (PTY) LTD
20 Alfred Street, Milsons Point, Sydney,
New South Wales 2061, Australia

RANDOM HOUSE NEW ZEALAND LTD
18 Poland Road, Glenfield, Auckland 10, New Zealand

RANDOM HOUSE SOUTH AFRICA (PTY) LTD
Isle of Houghton, Corner Boundary Road & Carse O'Gowrie,
Houghton 2198, South Africa.

Published 2005 by Bantam Press
a division of Transworld Publishers

Original art copyright 2005 © The Book Laboratory® Inc.
Image of Professor Stephen Hawking – Pages 21, 34 and 93 © Stewart Cohen
Cover Art – The Book Laboratory® Inc. and Moonrunner Design
Acknowledgements – Book Illustrations – The Book Laboratory® Inc.,
James Zhang and Kees Veenenbos
Image of Marilyn Monroe – The Estate of Andre de Dienes/Ms Shirley de Dienes
licenced by One West Publishing, Beverly Hills, Ca. 90212

Book design by Glen Edelstein

A catalogue record for this book is available
from the British Library
ISBN 0593 054970

Set in 12.5/15.5pt Lapidary by
Falcon Oast Graphic Art Ltd.

Printed and bound in Germany

1 3 5 7 9 10 8 6 4 2

Papers used by Transworld Publishers are natural, recyclable
products made from wood grown in sustainable forests. The
manufacturing processes conform to the environmental
regulations of the country of origin.

Contents

A BRIEFER HISTORY OF TIME

Acknowledgements

Thanks to our editor, Ann Harris, at Bantam for lending us her considerable experience and talent in our efforts to hone the manuscript. To Glen Edelstein, Bantam's art director, for his tireless efforts and his patience. To our art team, Philip Dunn, James Zhang, and Kees Veenenbos, for taking the time to learn some physics, and then, while not sacrificing the scientific content, making the book look fabulous. To our agents, Al Zuckerman and Susan Ginsburg at Writer's House, for their intelligence, caring, and support. To Monica Guy for proofreading. And to those who kindly read various drafts of the manuscript in our search for passages where clarity could be improved further: Donna Scott, Alexei Mlodinow, Nicolai Mlodinow, Mark Hillery, Joshua Webman, Stephen Youra, Robert Barkovitz, Martha Lowther, Katherine Ball, Amanda Bergen, Jeffrey Boehmer, Kimberly Comer, Peter Cook, Matthew Dickinson, Drew Donovanik, David Fralinger, Eleanor Grewal, Alicia Kingston, Victor Lamond, Michael Melton, Mychael Mulhern, Matthew Richards, Michelle Rose, Sarah Schmitt, Curtis Simmons, Christine Webb, and Christopher Wright.

Foreword

THE TITLE OF THIS BOOK DIFFERS by only two letters from that
of a book first published in 1988. *A Brief History of Time* was on the
London *Sunday Times* best-seller list for 237 weeks and has sold about
one copy for every 750 men, women, and children on earth. It was a
remarkable success for a book that addressed some of the most
difficult issues in modern physics. Yet those difficult issues are also
the most exciting, for they address big, basic questions: What do
we really know about the universe? How do we know it? Where did
the universe come from, and where is it going? Those questions
were the essence of *A Brief History of Time,* and they are also the focus
of this book.

In the years since *A Brief History of Time* was published, feedback
has come in from readers of all ages, of all professions, and from all
over the world. One repeated request has been for a new version, one
that maintains the essence of *A Brief History* yet explains the most important concepts in a clearer, more leisurely manner. Although one
might expect that such a book would be entitled *A Less Brief History of
Time,* it was also clear from the feedback that few readers are seeking
a lengthy dissertation befitting a college-level course in cosmology.

Thus, the present approach. In writing *A Briefer History of Time* we have maintained and expanded the essential content of the original book, yet taken care to maintain its length and readability. This is a briefer history indeed, for some of the more technical content has been left out, but we feel we have more than compensated for that by the more probing treatment of the material that is really the heart of the book.

We have also taken the opportunity to update the book and include new theoretical and observational results. *A Briefer History of Time* describes recent progress that has been made in finding a complete unified theory of all the forces of physics. In particular, it describes the progress made in string theory, and the "dualities" or correspondences between apparently different theories of physics that are an indication that there is a unified theory of physics. On the observational side, the book includes important new observations such as those made by the Cosmic Background Explorer satellite (COBE) and by the Hubble Space Telescope.

Some forty years ago Richard Feynman said, "We are lucky to live in an age in which we are still making discoveries. It is like the discovery of America—you only discover it once. The age in which we live is the age in which we are discovering the fundamental laws of nature." Today, we are closer than ever before to understanding the nature of the universe. Our goal in writing this book is to share some of the excitement of these discoveries, and the new picture of reality that is emerging as a result.

THINKING ABOUT
THE UNIVERSE

WE LIVE IN A STRANGE AND wonderful universe. Its age, size, violence, and beauty require extraordinary imagination to appreciate. The place we humans hold within this vast cosmos can seem pretty insignificant. And so we try to make sense of it all and to see how we fit in. Some decades ago, a well-known scientist (some say it was Bertrand Russell) gave a public lecture on astronomy. He described how the earth orbits around the sun and how the sun, in turn, orbits around the centre of a vast collection of stars called our galaxy. At the end of the lecture, a little old lady at the back of the room got up and said: "What you have told us is rubbish. The world is really a flat plate supported on the back of a giant turtle." The scientist gave a superior smile before replying, "What is the turtle standing on?" "You're very clever, young man, very clever," said the old lady. "But it's turtles all the way down!"

Most people nowadays would find the picture of our universe as an infinite tower of turtles rather ridiculous. But why should we think we know better? Forget for a minute what you know—or think you know—about space. Then gaze upwards at the night sky. What would you make of all those points of light? Are they tiny fires? It can be hard to imagine what they really are, for what they really are is far be-yond our ordinary experience. If you are a regular stargazer, you have probably seen an elusive light hovering near the horizon at twilight. It is a planet, Mercury, but it is nothing like our own planet. A day on Mercury lasts for two-thirds of the planet's year. Its surface reaches temperatures of over 400 degrees Celsius when the sun is out, then falls to almost −200 degrees Celsius in the dead of night. Yet as different as Mercury is from our own planet, it is not nearly as hard to imagine as a typical star, which is a huge furnace that burns billions of pounds of matter each second and reaches temperatures of tens of millions of degrees at its core.

Another thing that is hard to imagine is how far away the planets and stars really are. The ancient Chinese built stone towers so they could have a closer look at the stars. It's natural to think the stars and planets are much closer than they really are—after all, in everyday life we have no experience of the huge distances of space. Those distances are so large that it doesn't even make sense to measure them in feet or miles, the way we measure most lengths. Instead we use the light-year, which is the distance light travels in a year. In one second, a beam of light will travel 186,000 miles, so a light-year is a very long distance. The nearest star, other than our sun, is called Proxima Centauri (also known as Alpha Centauri C), which is about four light-years away. That is so far that even with the fastest spaceship on the drawing boards today, a trip to it would take about ten thousand years.

Ancient people tried hard to understand the universe, but they hadn't yet developed our mathematics and science. Today we have powerful tools: mental tools such as mathematics and the scientific method, and technological tools like computers and telescopes. With the help of these tools, scientists have pieced together a lot of knowl-edge about space. But what do we really know about the universe, and how do we know it? Where did the universe come from? Where is it going? Did the universe have a beginning, and if so, what happened before then? What is the nature of time? Will it ever come to an end? Can we go backwards in time? Recent breakthroughs in physics, made possible in part by new technology, suggest answers to some of these long-standing questions. Someday these answers may seem as obvious to us as the earth orbiting the sun—or perhaps as ridiculous as a tower of turtles. Only time (whatever that may be) will tell.

OUR EVOLVING PICTURE
OF THE UNIVERSE

ALTHOUGH AS LATE AS THE TIME of Christopher Columbus it was common to find people who thought the earth was flat (and you can even find a few such people today), we can trace the roots of modern astronomy back to the ancient Greeks. Around 340 B.C., the Greek philosopher Aristotle wrote a book called *On the Heavens*. In that book, Aristotle made good arguments for believing that the earth was a sphere rather than flat like a plate.

One argument was based on eclipses of the moon. Aristotle realized that these eclipses were caused by the earth coming between the sun and the moon. When that happened, the earth would cast its shadow on the moon, causing the eclipse. Aristotle noticed that the earth's shadow was always round. This is what you would expect if the earth was a sphere, but not if it was a flat disk. If the earth were a flat disk, its shadow would be round only if the eclipse happened at a time when the sun was directly under the centre of the disk. At other times the shadow would be elongated—in the shape of an ellipse (an ellipse is an elongated circle).

The Greeks had another argument for the earth being round. If the earth were flat, you would expect a ship approaching from the horizon to appear first as a tiny, featureless dot. Then, as it sailed closer, you would gradually be able to make out more detail, such as its sails and hull. But that is not what happens. When a ship appears on the horizon, the first things you see are the ship's sails. Only later do you see its hull. The fact that a ship's masts, rising high above the hull, are the first part of the ship to poke up over the horizon is evidence that the earth is a ball.

The Greeks also paid a lot of attention to the night sky. By Aristotle's time, people had for centuries been recording how the lights in the night sky moved. They noticed that although almost all of the thousands

Coming over the Horizon
Because the earth is a sphere, the mast and sails of a ship coming over the horizon show themselves before its hull.

of lights they saw seemed to move together across the sky, five of them (not counting the moon) did not. They would sometimes wander off from a regular east–west path and then double back. These lights were named planets—the Greek word for "wanderer". The Greeks observed only five planets because five are all we can see with the naked eye: Mercury, Venus, Mars, Jupiter, and Saturn. Today we know why the planets take such unusual paths across the sky: though the stars hardly move at all in comparison to our solar system, the planets orbit the sun, so their motion in the night sky is much more complicated than the motion of the distant stars.

Aristotle thought that the earth was stationary and that the sun, the moon, the planets, and the stars moved in circular orbits about the earth. He believed this because he felt, for mystical reasons, that the earth was the centre of the universe and that circular motion was the most perfect. In the second century A.D. another Greek, Ptolemy, turned this idea into a complete model of the heavens. Ptolemy was passionate about his studies. "When I follow at my pleasure the serried multitude of the stars in their circular course," he wrote, "my feet no longer touch the earth."

In Ptolemy's model, eight rotating spheres surrounded the earth. Each sphere was successively larger than the one before it, something like a Russian nesting doll. The earth was at the centre of the spheres. What lay beyond the last sphere was never made very clear, but it certainly was not part of mankind's observable universe. Thus the outermost sphere was a kind of boundary, or container, for the universe. The stars occupied fixed positions on that sphere, so when it rotated, the stars stayed in the same positions relative to each other and rotated together, as a group, across the sky, just as we observe. The inner spheres carried the planets. These were not fixed to their respective spheres as the stars were, but moved upon their spheres in small circles called epicycles. As the planetary spheres rotated and the planets themselves moved upon their spheres, the paths they took relative to the earth were complex ones. In this way, Ptolemy was able to account for the fact that the observed paths of the planets were much more complicated than simple circles across the sky.

Ptolemy's model provided a fairly accurate system for predicting the positions of heavenly bodies in the sky. But in order to predict these positions correctly, Ptolemy had to make an assumption that the moon followed a path that sometimes brought it twice as close to the earth as at other times. And that meant that the moon ought sometimes to appear twice as big as at other times! Ptolemy recognized this

Ptolemy's Model
In Ptolemy's model, the earth stood at the centre of the universe, surrounded by eight spheres carrying all the known heavenly bodies.

flaw, but nevertheless his model was generally, although not universally, accepted. It was adopted by the Christian church as the picture of the universe that was in accordance with scripture, for it had the great advantage that it left lots of room outside the sphere of fixed stars for heaven and hell.

Another model, however, was proposed in 1514 by a Polish priest, Nicolaus Copernicus. (At first, perhaps for fear of being branded a heretic by his church, Copernicus circulated his model anonymously.) Copernicus had the revolutionary idea that not all heavenly bodies must orbit the earth. In fact, his idea was that the sun was stationary at the centre of the solar system and that the earth and planets moved in circular orbits around the sun. Like Ptolemy's model, Copernicus's model worked well, but it did not perfectly match observation. Since it was much simpler than Ptolemy's model, though, one might have expected people to embrace it. Yet nearly a century passed before this idea was taken seriously. Then two astronomers—the German Johannes Kepler and the Italian Galileo Galilei—started publicly to support the Copernican theory.

In 1609, Galileo started observing the night sky with a telescope, which had just been invented. When he looked at the planet Jupiter, Galileo found that it was accompanied by several small satellites or moons that orbited around it. This implied that everything did not have to orbit directly around the earth, as Aristotle and Ptolemy had thought. At the same time, Kepler improved Copernicus's theory, suggesting that the planets moved not in circles but in ellipses. With this change the predictions of the theory suddenly matched the observations. These events were the death blows to Ptolemy's model.

Though elliptical orbits improved Copernicus's model, as far as Kepler was concerned they were merely a makeshift hypothesis. That is because Kepler had preconceived ideas about nature that were not based on any observation: like Aristotle, he simply believed that ellipses were less perfect than circles. The idea that planets would move along such imperfect paths struck him as too ugly to be the final truth. Another thing that bothered Kepler was that he could not make elliptical orbits consistent with his idea that the planets were made to orbit the sun by magnetic forces. Although he was wrong about magnetic forces being the reason for the planets' orbits, we have to give him credit for realizing that there must be a force responsible for the motion. The true explanation for why the planets orbit the sun was provided only much later, in 1687, when Sir Isaac Newton published his *Philosophiae Naturalis Principia Mathematica,* probably the most important single work ever published in the physical sciences.

In *Principia,* Newton presented a law stating that all objects at rest naturally stay at rest unless a force acts upon them, and described how the effects of force cause an object to move or change an object's motion. So why do the planets move in ellipses around the sun? Newton said that a particular force was responsible, and claimed that it was the same force that made objects fall to the earth rather than remain at rest when you let go of them. He named that force gravity (before Newton the word *gravity* meant only either a serious mood or a quality of heaviness). He also invented the mathematics that showed numerically how objects react when a force such as gravity pulls on them, and he solved the resulting equations. In this way he was able to show that due to the gravity of the sun, the earth and other planets should move in an ellipse—just as Kepler had predicted! Newton claimed that his laws applied to everything in the universe, from a falling apple to the stars and planets. It was the first time in history anybody had explained the motion of the planets in terms of laws that also determine motion on earth, and it was the beginning of both modern physics and modern astronomy.

Without the concept of Ptolemy's spheres, there was no longer any reason to assume the universe had a natural boundary, the outermost sphere. Moreover, since stars did not appear to change their positions apart from a rotation across the sky caused by the earth spinning on its axis, it became natural to suppose that the stars were objects like our sun but very much farther away. We had given up not only the idea that the earth is the centre of the universe but even the idea that our sun, and perhaps our solar system, were unique features of the cosmos. This change in worldview represented a profound transition in human thought: the beginning of our modern scientific understanding of the universe.

THE NATURE OF
A SCIENTIFIC THEORY

IN ORDER TO TALK ABOUT THE nature of the universe and to discuss such questions as whether it has a beginning or an end, you have to be clear about what a scientific theory is. We shall take the simpleminded view that a theory is just a model of the universe, or a restricted part of it, and a set of rules that relate quantities in the model to observations that we make. It exists only in our minds and does not have any other reality (whatever that might mean). A theory is a good theory if it satisfies two requirements. It must accurately describe a large class of observations on the basis of a model that contains only a few arbitrary elements, and it must make definite predictions about the results of future observations. For example, Aristotle believed Empedocles' theory that everything was made out of four elements: earth, air, fire, and water. This was simple enough but did not make any definite predictions. On the other hand, Newton's theory of gravity was based on an even simpler model, in which bodies attracted each other with a force that was proportional to a quantity called their mass and inversely proportional to the square of the distance between them. Yet it predicts the motions of the sun, the moon, and the planets to a high degree of accuracy.

Any physical theory is always provisional, in the sense that it is only a hypothesis: you can never prove it. No matter how many times the results of experiments agree with some theory, you can never be sure that the next time a result will not contradict the theory. On the other hand, you can disprove a theory by finding even a single observation that disagrees with the predictions of the theory. As philosopher of science Karl Popper has emphasized, a good theory is characterized by the fact that it makes a number of predictions that could in principle be disproved or falsified by observation. Each time new experiments are observed to agree with the predictions, the theory survives and our confidence in it is increased; but if ever a new observation is found to disagree, we have to abandon or modify the theory.

At least that is what is supposed to happen, but you can always question the competence of the person who carried out the observation.

In practice, what often happens is that a new theory is devised that is really an extension of the previous theory. For example, very accurate observations of the planet Mercury revealed a small difference between its motion and the predictions of Newton's theory of gravity. Einstein's general theory of relativity predicted a slightly different motion than Newton's theory did. The fact that Einstein's predictions matched what was seen, while Newton's did not, was one of the crucial confirmations of the new theory. However, we still use Newton's theory for most practical purposes because the difference between its predictions and those of general relativity is very small in the situations that we normally deal with. (Newton's theory also has the great advantage that it is much simpler to work with than Einstein's!)

The eventual goal of science is to provide a single theory that describes the whole universe. However, the approach most scientists actually follow is to separate the problem into two parts. First, there are the laws that tell us how the universe changes with time. (If we know what the universe is like at any one time, these physical laws tell us how it will look at any later time.) Second, there is the question of the initial state of the universe. Some people feel that science should be concerned with only the first part; they regard the question of the initial situation as a matter for metaphysics or religion. They would say that God, being omnipotent, could have started the universe off any way He wanted. That may be so, but in that case God also could have made it develop in a completely arbitrary way. Yet it appears that God chose to make it evolve in a very regular way, according to certain laws. It therefore seems equally reasonable to suppose that there are also laws governing the initial state.

It turns out to be very difficult to devise a theory to describe the universe all in one go. Instead, we break the problem up into bits and invent a number of partial theories. Each of these partial theories describes and predicts a certain limited class of observations, neglecting the effects of other quantities, or representing them by simple sets of numbers. It may be that this approach is completely wrong. If everything in the universe depends on everything else in a fundamental way, it might be impossible to get close to a full solution by investigating parts of the problem in isolation. Nevertheless, it is certainly the way that we have made progress in the past. The classic example is again the Newtonian theory of gravity, which tells us that the gravitational force between two bodies depends only on one number associated with each body, its mass, and is otherwise independent of what the bodies are made of. Thus we do not need to have a theory of the structure and constitution of the sun and the planets in order to calculate their orbits.

Today scientists describe the universe in terms of two basic partial theories—the general theory of relativity and quantum mechanics. They are the great intellectual achievements of the first half of the twentieth century. The general theory of relativity describes the force of gravity and the large-scale structure of the universe; that is, the structure on scales from only a few miles to as large as a million million million million (1 with twenty-four zeros after it) miles, the size of the observable universe. Quantum mechanics, on the other hand,

Atoms to Galaxies
In the first half of the twentieth century, physicists extended the reach of their theories from the everyday world of Isaac Newton to both the smallest and the largest extremes of our universe.

deals with phenomena on extremely small scales, such as a millionth of a millionth of an inch. Unfortunately, however, these two theories are known to be inconsistent with each other—they cannot both be correct. One of the major endeavours in physics today, and the major theme of this book, is the search for a new theory that will incorporate them both—a quantum theory of gravity. We do not yet have such a theory, and we may still be a long way from having one, but we do already know many of the properties that it must have. And we shall see in later chapters that we already know a fair amount about the predictions a quantum theory of gravity must make.

Now, if you believe that the universe is not arbitrary but is governed by definite laws, you ultimately have to combine the partial theories into a complete unified theory that will describe everything in the universe. But there is a fundamental paradox in the search for such a complete unified theory. The ideas about scientific theories outlined above assume we are rational beings who are free to observe the universe as we want and to draw logical deductions from what we see. In such a scheme it is reasonable to suppose that we might progress ever closer towards the laws that govern our universe. Yet if there really were a complete unified theory, it would also presumably determine our actions—so the theory itself would determine the outcome of our search for it! And why should it determine that we come to the right conclusions from the evidence? Might it not equally well determine that we draw the wrong conclusion? Or no conclusion at all?

The only answer that we can give to this problem is based on Darwin's principle of natural selection. The idea is that in any population of self-reproducing organisms, there will be variations in the genetic material and upbringing that different individuals have. These differences will mean that some individuals are better able than others to draw the right conclusions about the world around them and to act accordingly. These individuals will be more likely to survive and reproduce, so their pattern of behaviour and thought will come to dominate. It has certainly been true in the past that what we call intelligence and scientific discovery have conveyed a survival advantage. It is not so clear that this is still the case: our scientific discoveries may well destroy us all, and even if they don't, a complete unified theory may not make much difference to our chances of survival. However, provided the universe has evolved in a regular way, we might expect that the reasoning abilities that natural selection has given us would also be valid in our search for a complete unified theory and so would not lead us to the wrong conclusions.

Because the partial theories that we already have are sufficient to make accurate predictions in all but the most extreme situations, the search for the ultimate theory of the universe seems difficult to justify on practical grounds. (It is worth noting, though, that similar arguments could have been used against both relativity and quantum mechanics, and these theories have given us both nuclear energy and the microelectronics revolution.) The discovery of a complete unified theory, therefore, may not aid the survival of our species. It may not even affect our lifestyle. But ever since the dawn of civilization, people have not been content to see events as unconnected and inexplicable. We have craved an understanding of the underlying order in the world. Today we still yearn to know why we are here and where we came from. Humanity's deepest desire for knowledge is justification enough for our continuing quest. And our goal is nothing less than a complete description of the universe we live in.

• 4 •

NEWTON'S UNIVERSE

OUR PRESENT IDEAS ABOUT THE MOTION of bodies date back to Galileo and Newton. Before them, people believed Aristotle, who said that the natural state of a body was to be at rest, and that it moved only if driven by a force or impulse. It followed that a heavier body should fall faster than a light one because it would have a greater pull towards the earth. The Aristotelian tradition also held that one could work out all the laws that govern the universe by pure thought: it was not necessary to check by observation. So no one until Galileo bothered to see whether bodies of different weights did in fact fall at different speeds. It is said that Galileo demonstrated that Aristotle's belief was false by dropping weights from the Leaning Tower of Pisa in Italy. This story is almost certainly untrue, but Galileo did do something equivalent: he rolled balls of different weights down a smooth slope. The situation is similar to that of heavy bodies falling vertically, but it is easier to observe because the speeds are smaller. Galileo's measurements indicated that each body increased its speed at the same rate, no matter what its weight. For example, if you let go of a ball on a slope that drops by one metre for every ten metres you go along, the ball will be travelling down the slope at a speed of about one metre per second after one second, two metres per second after two seconds, and so on, however heavy the ball. Of course a lead weight would fall faster than a feather, but that is only because a feather is slowed down by air resistance. If you drop two bodies that don't have much air resistance, such as two different lead weights, they fall at the same rate. (We will see why shortly.) On the moon, where there is no air to slow things down, the astronaut David R. Scott performed the feather-and-lead-weight experiment and found that indeed they did hit the ground at the same time.

Galileo's measurements were used by Newton as the basis of his laws of motion. In Galileo's experiments, as a body rolled down the slope it was always acted on by the same force (its weight), and the effect was to make it constantly speed up. This showed that the real effect of a force is always to change the speed of a body, rather than just to set it moving, as was previously thought. It also meant that whenever a body is not acted on by any force, it will keep on moving in a straight line at the same speed. This idea was first stated explicitly in 1687, in Newton's *Principia Mathematica,* and is known as Newton's first law. What happens to a body when a force does act on it is given by Newton's second law. This states that the body will accelerate, or change its speed, at a rate that is proportional to the force. (For example, the acceleration is twice as great if the force is twice as great.) The acceleration is also smaller the greater the mass (or quantity of matter) of the body. (The same force acting on a body of twice the mass will produce half the acceleration.) A familiar example is provided by a car: the more powerful the engine, the greater the acceleration, but the heavier the car, the smaller the acceleration for the same engine.

In addition to his laws of motion, which describe how bodies react to forces, Newton's theory of gravity describes how to determine the strength of one particular type of force, gravity. As we have said, that theory states that every body attracts every other body with a force that is proportional to the mass of each body. Thus, the force between two bodies would be twice as strong if one of the bodies (say, body A) had its mass doubled. This is what you might expect, because one could think of the new body A as being made of two bodies, each with the original mass. Each of these would attract body B with the original force. Thus the total force between A and B would be twice the original force. And if, say, one of the bodies had six times the mass, or one had twice the mass and the other had three times the mass, then the force between them would be six times as strong.

You can now see why all bodies fall at the same rate. According to Newton's law of gravity, a body of twice the weight will have twice the

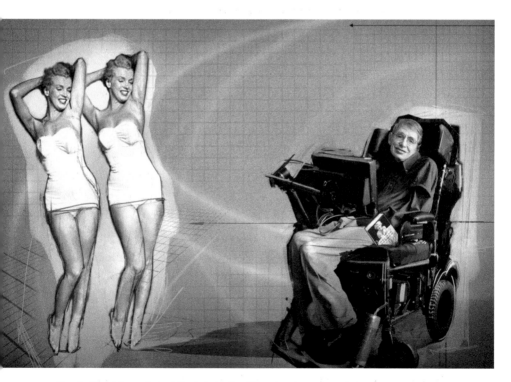

Gravitational Attraction of Composite Bodies
If the mass of a body is doubled, so is the gravitational force that it exerts.

force of gravity pulling it down. But it will also have twice the mass and thus, according to Newton's second law, half the acceleration per unit force. According to Newton's laws, these two effects exactly cancel each other out, so the acceleration will be the same no matter what the weight.

Newton's law of gravity also tells us that the farther apart the bodies, the lesser the force. The law says that the gravitational attraction of a star is exactly one-quarter that of a similar star at half the distance. This law predicts the orbits of the earth, the moon, and the planets with great accuracy. If the law were that the gravitational attraction of a star went down faster or slower with distance, the orbits of the planets would not be elliptical; they would either spiral into the sun or escape from the sun.

The big difference between the ideas of Aristotle and those of Galileo and Newton is that Aristotle believed in a preferred state of rest, which any body would take up if it was not driven by some force or impulse. In particular, he thought that the earth was at rest. But it follows from Newton's laws that there is no unique standard of rest. One could equally well say that body A was at rest and body B was moving at constant speed with respect to body A, or that body B was at rest and body A was moving. For example, if you set aside for a moment the rotation of the earth and its orbit around the sun, you could say that the earth was at rest and that a train on it was travelling north at ninety miles per hour or that the train was at rest and the earth was moving south at ninety miles per hour. If you carried out experiments with moving bodies on the train, all Newton's laws would still hold. Is Newton right or is Aristotle, and how do you tell?

One test would be this: imagine you are enclosed in a box, and you do not know whether the box is resting on the floor of a moving train or on solid earth, the latter being the standard of rest according to Aristotle. Is there a way to determine which it is? If so, maybe Aristotle was correct—being at rest on the earth is special. But if you carried out experiments in your box on the train, they would turn out exactly the same as they would within your box on the "stationary" train platform (assuming no bumps, turns, or other imperfections in the train ride). Playing Ping-Pong on the train, you would find that the ball behaved just like a ball on a Ping-Pong table by the track. And if you are in your box and play the game at different speeds relative to the earth, say at zero, fifty, and ninety miles per hour, the ball will behave the same in all these situations. This is how the world behaves, and it is what the mathematics of Newton's laws reflects: there is no way to tell whether it is the train or the earth that is moving. The concept of motion makes sense only as it relates to other objects.

Does it really matter whether Aristotle or Newton is correct? Is this merely a difference in outlook or philosophy, or is it an issue important to science? Actually, the lack of an absolute standard of rest has deep implications for physics: it means that we cannot determine whether two events that took place at different times occurred in the same position in space.

To picture this, suppose someone on a train bounces a Ping-Pong ball straight up and down, hitting the table twice on the same spot one second apart. To that person, the locations of the first and second bounces will have a spatial separation of zero. To someone standing beside the track, the two bounces would seem to take place about forty metres apart, because the train would have travelled that distance down the track between the bounces. According to Newton, the two observers have an equal right to consider themselves at rest, so both views are equally acceptable. One is not favoured over another, as Aristotle had believed. The observed positions of events and the distances between them would be different for a person on the train and one beside the track, and there would be no reason to prefer one person's observations to the other's.

Newton was very worried by this lack of absolute position, or absolute space, as it was called, because it did not accord with his idea of an absolute God. In fact, he refused to accept the lack of absolute space,

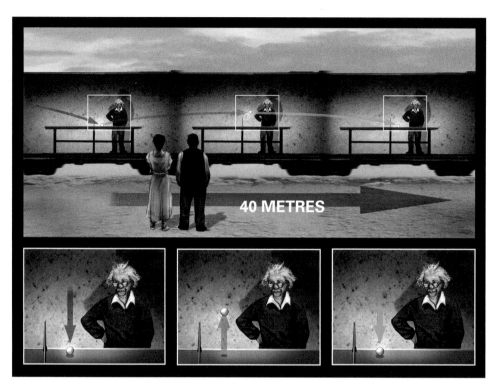

Relativity of Distance
The distance—and the path—that an object travels can look different to different observers.

even though his laws implied it. He was severely criticized for this irrational belief by many people, most notably by Bishop Berkeley, a philosopher who believed that all material objects and space and time are an illusion. When the famous Dr. Johnson was told of Berkeley's opinion, he cried, "I refute it thus!" and stubbed his toe on a large stone.

Both Aristotle and Newton believed in absolute time. That is, they believed that one could unambiguously measure the interval of time between two events and that this time would be the same whoever measured it, provided the person used a good clock. Unlike absolute space, absolute time was consistent with Newton's laws.

And it is what most people would take to be the commonsense view. However, in the twentieth century physicists realized that they had to change their ideas about both time and space. As we will see, they discovered that the length of time between events, like the distance between the points where the Ping-Pong ball bounced, depends on the observer. They also discovered that time was not completely separate from and independent of space. The key to these realizations was new insight into the properties of light. They may seem counter to our experience, but although our apparently commonsense notions work well when dealing with things such as apples, or planets that travel comparatively slowly, they don't work at all for things moving at or near the speed of light.

RELATIVITY

THE FACT THAT LIGHT TRAVELS AT a finite but very high speed was first discovered in 1676 by the Danish astronomer Ole Christensen Roemer. If you observe the moons of Jupiter, you will notice that from time to time they disappear from sight because they pass behind the giant planet. These eclipses of Jupiter's moons ought to occur at regular intervals, but Roemer observed that the eclipses were not evenly spaced. Did the moons somehow speed up and slow down in their orbits? He had another explanation. If light travelled with infinite speed, then we on earth would see the eclipses at regular intervals, at exactly the same time that they occurred, like the ticks of a cosmic clock. Since light would traverse any distance instantaneously, this situation would not change if Jupiter moved closer to the earth or further from it.

Now imagine that light travels with finite speed. If so, we will see each eclipse some time after it happened. This delay depends upon the speed of light *and* on the distance of Jupiter from the earth. If Jupiter did not change its distance from the earth, the delay would be the same for every eclipse. However, Jupiter sometimes moves closer to the earth. In such cases, the "signal" from each successive eclipse has less and less distance to travel, so it arrives progressively earlier than if

Jupiter had remained at a constant distance. For analogous reasons, when Jupiter is receding from the earth, we see the eclipses progressively later. The degree of this early and late arrival depends upon the speed of light, and this allows us to measure it. This is what Roemer did. He noticed that eclipses of one of Jupiter's moons appeared earlier at those times of year when the earth was approaching Jupiter's orbit and later at those times when the earth was moving away, and he used this

The Speed of Light and the Timing of Eclipses
The observed times of the eclipses of Jupiter's moons depend on both the actual time of the eclipses and the time it takes their light to travel from Jupiter to the earth. Thus the eclipses seem to appear more frequently when Jupiter is moving towards the earth, and less frequently when it is moving away. This effect is exaggerated here for clarity.

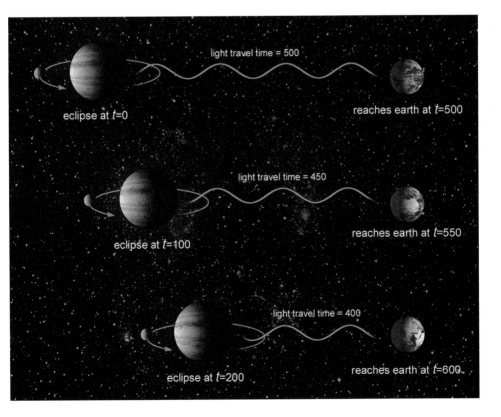

difference to calculate the speed of light. However, his measurements of the variations in the distance of the earth from Jupiter were not very accurate, so his value for the speed of light was 140,000 miles per second, compared to the modern value of 186,000 miles per second. Nevertheless, Roemer's achievement, not only in proving that light travels at a finite speed but also in measuring that speed, was remarkable, coming as it did eleven years before Newton's publication of *Principia Mathematica.*

A proper theory of the propagation of light didn't come until 1865, when the British physicist James Clerk Maxwell succeeded in unifying the partial theories that up to then had been used to describe the forces of electricity and magnetism. Though both electricity and magnetism had been known since ancient times, it wasn't until the eighteenth century that British chemist Henry Cavendish and French physicist Charles-Augustin de Coulomb established quantitative laws governing the electric force between two charged bodies. A few decades later, in the early nineteenth century, a number of physicists established analogous laws for magnetic forces. Maxwell showed mathematically that these electric and magnetic forces do not arise from particles acting directly on each other; rather, every electric charge and current creates a field in the surrounding space that exerts a force on every other charge and current located within that space. He found that a single field carries the electric and magnetic forces; thus, electricity and magnetism are inseparable aspects of the same force. He called that force the electromagnetic force, and the field that carries it the electromagnetic field.

Maxwell's equations predicted that there could be wavelike disturbances in the electromagnetic field and that these waves would travel at a fixed speed, like ripples on a pond. When he calculated this speed, he found it to match exactly the speed of light! Today we know that Maxwell's waves are visible to the human eye as light when they have a wavelength of between forty and eighty millionths of a centimetre. (A wave is a succession of crests and troughs; the wavelength is the distance between wave crests or troughs.) Waves with wavelengths shorter than those of visible light are now known as

Wavelength
The wavelength of a wave is the distance between successive peaks or troughs.

ultraviolet light, X-rays, and gamma rays. Waves with longer wave-lengths are called radio waves (a metre or more), microwaves (around a centimetre), or infrared radiation (less than one ten-thousandth of a centimetre but more than the visible range).

Maxwell's theory implied that radio or light waves would travel at a certain fixed speed. This was difficult to reconcile with Newton's theory that there is no absolute standard of rest, because if there is no such standard, there can be no universal agreement on the speed of an object. To understand why, again imagine yourself playing Ping-Pong on the train. If you hit the ball towards the front of the train with a speed your opponent measures to be ten miles per hour, then you'd expect an observer on the platform to perceive the ball moving at one hundred miles per hour—the ten it is moving relative to the train, plus the ninety the train is moving relative to the platform. What is the speed of the ball, ten miles per hour or one hundred? How do you define it— relative to the train or relative to the earth? With no absolute standard

Different Speeds of Ping-Pong Balls
According to the theory of relativity, although they may disagree, every observer's measurement of an object's speed is equally valid.

of rest, you cannot assign the ball an absolute speed. The same ball could equally well be said to have any speed, depending upon the frame of reference in which the speed is measured. According to Newton's theory, the same should hold for light. So what does it mean in Maxwell's theory for light waves to travel at a certain fixed speed?

In order to reconcile Maxwell's theory with Newton's laws, it was suggested that there was a substance called the ether that was present everywhere, even in the vacuum of "empty" space. The idea of the ether had a certain added attraction for scientists who felt in any case that, just as water waves require water or sound waves require air, waves of electromagnetic energy must require some medium to carry them. In this view, light waves travel through the ether as sound waves travel through air, and their "speed" as derived from Maxwell's equations should therefore be measured relative to the ether. Different observers would see light coming towards them at different speeds, but light's speed relative to the ether would remain fixed.

This idea could be tested. Imagine light emitted from some source. According to the ether theory, the light travels through the ether at the speed of light. If you move towards it through the ether, the speed at which you approach the light will be the sum of the speed of light through the ether and your speed through the ether. The light will approach you faster than if, say, you didn't move, or you moved in some other direction. Yet because the speed of light is so great compared to the speeds at which we might move towards a light source, this difference in speed was a very difficult effect to measure.

In 1887, Albert Michelson (who later became the first American to receive the Nobel Prize for physics) and Edward Morley carried out a very careful and difficult experiment at the Case School of Applied Science (now Case Western Reserve University) in Cleveland. They realized that because the earth orbits the sun at a speed of nearly twenty miles per second, their lab itself must be moving at a relatively high rate of speed through the ether. Of course, no one knew in which direction or how fast the ether might be moving with respect to the sun, or whether it was moving at all. But by repeating an experiment at different times of the year, when the earth was in different positions along its orbit, they could hope to account for this unknown factor. So Michelson and Morley set up an experiment to compare the speed of light measured in the direction of the earth's motion through the ether (when we were moving towards the source of the light) to the speed of light at right angles to that motion (when we were not moving towards the source). To their great surprise, they found the speed in both directions was exactly the same!

Between 1887 and 1905, there were several attempts to save the ether theory. The most notable was by the Dutch physicist Hendrik Lorentz, who attempted to explain the result of the Michelson–Morley experiment in terms of objects contracting and clocks slowing down when they moved through the ether. However, in a famous paper in 1905, a hitherto unknown clerk in the Swiss patent office, Albert Einstein, pointed out that the whole idea of an ether was unnecessary, provided one was willing to abandon the idea of absolute time (we'll see why shortly). A leading French mathematician, Henri Poincaré, made a similar point a few weeks later. Einstein's arguments were closer to physics than those of Poincaré, who regarded this problem as purely mathematical and to his dying day did not accept Einstein's interpretation of the theory.

Einstein's fundamental postulate of the theory of relativity, as it was called, stated that the laws of science should be the same for all freely moving observers, no matter what their speed. This was true for Newton's laws of motion, but now Einstein extended the idea to include Maxwell's theory. In other words, since Maxwell's theory dictates that the speed of light has a given value, all freely moving observers must measure that same value, no matter how fast they are moving towards or away from its source. This simple idea certainly explained—without the use of the ether or any other preferred frame of reference—the meaning of the speed of light in Maxwell's equations, yet it also had some remarkable and often counterintuitive consequences.

For example, the requirement that all observers must agree on how fast light travels forces us to change our concept of time. Picture again the speeding train. In Chapter 4, we saw that although someone on the train bouncing a Ping-Pong ball up and down may say that the ball travelled only a few inches, someone standing on the platform would perceive the ball as travelling about forty metres. Similarly, if the observer on the train shone a torch, the two observers would disagree on the distance the light travelled. Since speed is distance divided by time, if they disagree on the distance the light has travelled, the only way for them to agree on the speed of light is for them to also disagree about the time the trip has taken. In other words, the theory of relativity requires us to put an end to the idea of absolute time! Instead, each observer must have his own measure of time, as recorded by a clock carried with him, and identical clocks carried by different observers need not agree.

In relativity there is no need to introduce the idea of an ether, whose presence, as the Michelson–Morley experiment showed, cannot be detected. Instead, the theory of relativity forces us to change fundamentally our ideas of space and time. We must accept that time is not completely separate from and independent of space but is combined with it to form an object called space-time. These are not easy ideas to grasp. Relativity took years to become universally accepted even within the physics community. It is a testament to Einstein's imagination that he was able to conceive it, and to his confidence in his own logic that he worked out its consequences despite the odd conclusions towards which it seemed to be leading.

It is a matter of common experience that we can describe the position of a point in space by three numbers, or coordinates. For instance, we can say that a point in a room is seven metres from one wall, three metres from another, and five metres above the floor. Or we could specify that a point is at a certain latitude and longitude and a certain height above sea level. We are free to use any three suitable

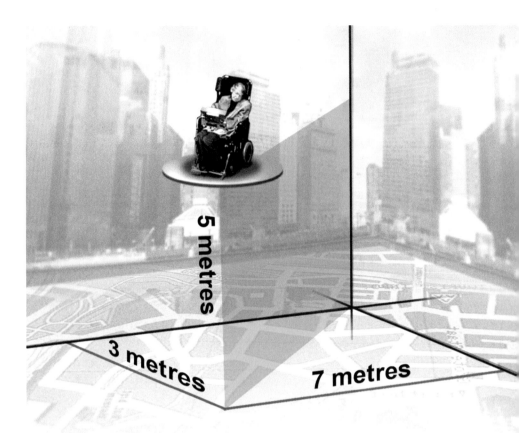

Coordinates in Space
When we say that space has three dimensions, we mean that it takes three numbers, or *coordinates,* to specify a point. If we add time to our description, then space becomes space-time, with four dimensions.

coordinates, although they have only a limited range of validity. It would not be practical to specify the position of the moon in terms of miles north and miles west of Piccadilly Circus and feet above sea level. Instead, we might describe it in terms of distance from the sun, distance from the plane of the orbits of the planets, and the angle between the line joining the moon to the sun and the line joining the sun to a nearby star such as Proxima Centauri. Even these coordinates would not be of much use in describing the position of the sun in our galaxy or the position of our galaxy in the local group of galaxies. In fact, we may describe the whole universe in terms of a collection of overlapping patches. In each patch, we can use a different set of three coordinates to specify the position of a point.

In the space-time of relativity, any event—that is, anything that happens at a particular point in space and at a particular time—can be specified by *four* numbers or coordinates. Again, the choice of co-ordinates is arbitrary: we can use any three well-defined spatial coor-dinates and any measure of time. But in relativity, there is no real distinction between the space and time coordinates, just as there is no real difference between any two space coordinates. We could choose a new set of coordinates in which, say, the first space coordinate was a combination of the old first and second space coordinates. So instead of measuring the position of a point on the earth in miles north of Piccadilly and miles west of Piccadilly, we could use miles north-east of Piccadilly and miles north-west of Piccadilly. Similarly, we could use a new time coordinate that was the old time (in seconds) plus the distance (in light-seconds) north of Piccadilly.

Another well-known consequence of relativity is the equiva-lence of mass and energy, summed up in Einstein's famous equation $E = mc^2$ (where E is energy, m is mass, and c is the speed of light). People often employ this equation to calculate how much energy would be produced if, say, a bit of matter was converted into pure electromagnetic radiation. (Because the speed of light is a large number, the answer is a lot—the weight of matter converted to energy in the bomb that destroyed the city of Hiroshima was less than one ounce.) But the equation also tells us that if the energy of an object increases, so does its mass, that is, its resistance to accelera-tion, or change in speed.

One form of energy is energy of motion, called kinetic energy. Just as it takes energy to get your car moving, it takes energy to in-crease the speed of any object. The kinetic energy of a moving object is identical to the energy you must expend in causing it to move. Therefore, the faster an object moves, the more kinetic energy it possesses. But according to the equivalence of energy and mass, kinetic energy adds to an object's mass, so the faster an object moves, the harder it is to further increase the object's speed.

This effect is really significant only for objects moving at speeds close to the speed of light. For example, at 10 per cent of the speed of light, an object's mass is only 0.5 per cent more than normal, while at 90 per cent of the speed of light it would be more than twice its normal mass. As an object approaches the speed of light, its mass rises ever more quickly, so it takes more and more energy to speed it up further. According to the theory of relativity, an object can in fact never reach the speed of light, because by then its mass would have become infinite, and by the equivalence of mass and energy, it would have taken an infinite amount of energy to get it there. This is the reason that any normal object is forever confined by relativity to move at speeds slower than the speed of light. Only light, or other waves that have no intrinsic mass, can move at the speed of light.

Einstein's 1905 theory of relativity is called special relativity. That is because, though it was very successful in explaining that the speed of light was the same to all observers and in explaining what happens when things move at speeds close to the speed of light, it was inconsistent with the Newtonian theory of gravity. Newton's theory says that at any given time objects are attracted to each other with a force that depends on the distance between them at that time. This means that if you moved one of the objects, the force on the other one would change instantaneously. If, say, the sun suddenly disappeared, Maxwell's theory tells us that the earth wouldn't get dark for about another eight minutes (since that is how long it takes light to reach us from the sun) but, according to Newtonian gravity, the earth would immediately cease to feel the sun's attraction and fly out of orbit. The gravitational effect of the disappearance of the sun would thus have reached us with infinite speed, instead of at or below the speed of light, as the special theory of relativity required. Einstein made a number of unsuccessful attempts between 1908 and 1914 to find a theory of gravity that was consistent with special relativity. Finally, in 1915, he proposed the even more revolutionary theory we now call the general theory of relativity.

· 6 ·

CURVED SPACE

EINSTEIN'S THEORY OF GENERAL RELATIVITY IS based on the revolutionary suggestion that gravity is not a force like other forces but a consequence of the fact that space-time is not flat, as had been previously assumed. In general relativity, space-time is curved, or "warped", by the distribution of mass and energy in it. Bodies such as the earth are not made to move on curved orbits by a force called gravity; instead they move in curved orbits because they follow the nearest thing to a straight path in a curved space, which is called a geodesic. Technically speaking, a geodesic is defined as the shortest (or longest) path between two nearby points.

A geometric plane is an example of a two-dimensional flat space, on which the geodesics are lines. The surface of the earth is a two-dimensional curved space. A geodesic on the earth is called a great circle. The equator is a great circle. So is any other circle on the globe whose centre coincides with the centre of the earth. (The term "great circle" comes from the fact that these are the largest circles you can draw on the globe.) As the geodesic is the shortest path between two airports, this is the route an airline navigator will tell the pilot to fly along. For instance, you could fly from New York to Madrid by following your compass for 3,707 miles almost straight east, along their common line of latitude. But you can get there in 3,605 miles if you fly along a great circle, heading first north-east, then gradually turning east, and then south-east. The appearance of these two paths on a

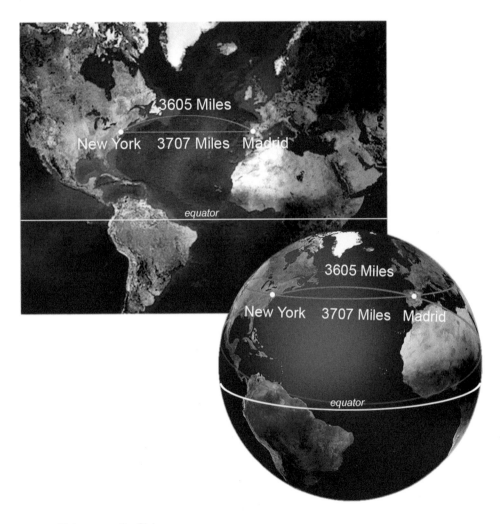

Distances on the Globe
The shortest distance between two points on the globe is along a great circle, which does not correspond to a straight line if you are looking at a flat map.

map, in which the surface of the globe has been distorted (flattened out), is deceiving. When you move "straight" east, you are not really moving straight, at least not straight in the sense of the most direct path, the geodesic.

In general relativity, bodies always follow geodesics in four-dimensional space-time. In the absence of matter, these geodesics in four-dimensional space-time correspond to straight lines in three-

Path of a Spacecraft's Shadow
Projected on to the two-dimensional globe, the path of a spacecraft flying along a straight line in space will appear curved.

dimensional space. In the presence of matter, four-dimensional space-time is distorted, causing the paths of bodies in three-dimensional space to curve in a manner that in the old Newtonian theory was explained by the effects of gravitational attraction. This is rather like watching an airplane flying over hilly ground. The plane might be moving in a straight line through three-dimensional space, but remove the third dimension—height—and you find that its shadow follows a curved path on the hilly two-dimensional ground. Or imagine a spaceship flying in a straight line through space, passing directly over the North Pole. Project its path down onto the two-dimensional surface of the earth and you find that it follows a semicircle, tracing a line of longitude over the northern hemisphere. Though the phenomenon is harder to picture, the mass of the sun curves space-time in such a way that although the earth follows a straight path in four-dimensional space-time, it appears to us to move along a nearly circular orbit in three-dimensional space.

Actually, although they are derived differently, the orbits of the planets predicted by general relativity are almost exactly the same as those predicted by the Newtonian theory of gravity. The largest deviation is in the orbit of Mercury, which, being the planet nearest to the sun, feels the strongest gravitational effects and has a rather elongated elliptical orbit. General relativity predicts that the long axis of the ellipse should rotate about the sun at a rate of about one degree per ten thousand years. Small though this effect is, it had been noticed (see Chapter 3) long before 1915, and it served as one of the first confirmations of Einstein's theory. In recent years, the even smaller deviations of the orbits of the other planets from the Newtonian predictions have been measured by radar and found to agree with the predictions of general relativity.

Light rays too must follow geodesics in space-time. Again, the fact that space is curved means that light no longer appears to travel in

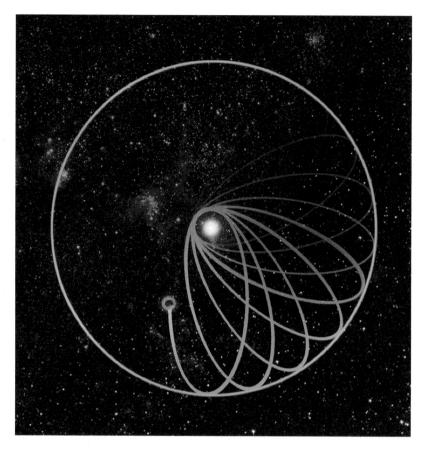

Precession of Mercury's Orbit
As Mercury repeatedly orbits the sun, the long axis of its elliptical path slowly rotates, coming full circle roughly every 360,000 years.

straight lines in space, so general relativity predicts that gravitational fields should bend light. For example, the theory predicts that the path of light near the sun would be slightly bent inwards, on account of the mass of the sun. This means that light from a distant star that happened to pass near the sun would be deflected through a small angle, causing the star to appear in a different position to an observer on the earth. Of course, if the light from the star always passed close

Bending of Light Near the Sun

When the sun lies almost directly between the earth and a distant star, its gravitational field deflects the star's light, altering its apparent position.

to the sun, we would not be able to tell whether the light was being deflected or if instead the star was really where we seem to see it. However, as the earth orbits around the sun, different stars appear to pass behind the sun and have their light deflected. They therefore change their apparent position relative to the other stars.

It is normally very difficult to see this effect, because the light from the sun makes it impossible to observe stars that appear near to the sun in the sky. However, it is possible to do so during an eclipse of the sun, when the moon blocks the sun's light. Einstein's prediction of light deflection could not be tested immediately in 1915, because the First World War was in progress. It was not until 1919 that a British expedition, observing an eclipse from the coast of West Africa, showed that light was indeed deflected by the sun, just as predicted by the theory. This proof of a German theory by British scientists was hailed as a great act of reconciliation between the two countries after the war. It is ironic, therefore, that later examination of the photographs taken on that expedition showed the errors were as great as the effect they were trying to measure. Their measurement had been sheer luck, or perhaps a case of knowing the result they wanted to get—not an uncommon occurrence in science. The light deflection has, however, been accurately confirmed by a number of later observations.

Another prediction of general relativity is that time should appear to run slower near a massive body such as the earth. Einstein first realized this in 1907, five years before he realized that gravity also altered the shape of space, and eight years before he completed his theory. He derived the effect using his principle of equivalence, which played the role in general relativity that the fundamental postulate played in the special theory.

Recall that the fundamental postulate of special relativity stated that the laws of science should be the same for all freely moving observers, no matter what speed they were moving at. Roughly speaking, the principle of equivalence extends this to those observers who are not freely moving but are under the influence of a gravitational field. In a precise statement of the principle there are some technical points, such as the fact that if the gravitational field is not uniform, you must apply the principle separately to a series of small, overlapping patches, but we won't concern ourselves with that here. For our purposes, we can state the principle this way: in small enough regions of space, it is impossible to tell if you are at rest in a gravitational field or uniformly accelerating in empty space.

Imagine that you are in a lift in empty space. There is no gravity, no "up" and no "down." You are floating freely. Now the lift starts to move with constant acceleration. You suddenly feel weight. That is, you feel a pull towards one end of the lift, which suddenly seems to you to be the floor! If you now hold out an apple and let go, it drops to the floor. In fact, now that you are accelerating, everything that happens inside the lift will unfold exactly as it would if the lift was not moving at all but at rest in a uniform gravitational field. Einstein realized that just as you cannot tell from inside a train whether or not you are moving uniformly, you also cannot tell from inside the lift whether you are uniformly accelerating or in a uniform gravitational field. The result was his principle of equivalence.

The principle of equivalence, and the above example of it, is true only if inertial mass (the mass in Newton's second law that determines how much you accelerate in response to a force) and gravitational mass (the mass in Newton's law of gravity that determines how much gravitational force you feel) are the same thing (see Chapter 4). That's because if both kinds of mass are the same, then all objects in a gravitational field will fall at the same rate, no matter what their mass. If this equivalence weren't true, then under the influence of gravity some objects would fall faster than others, which would mean you could distinguish the pull of gravity from uniform acceleration, in which everything does fall at the same rate. Einstein's use of the equivalence of inertial and gravitational mass to derive his principle of equivalence, and eventually all of general relativity, amounts to a relentless march of logical reasoning unmatched in the history of human thought.

Now that we know the principle of equivalence, we can start to follow Einstein's logic by doing another thought experiment that shows why time must be affected by gravity. Imagine a rocket ship out in space. For convenience, imagine that the rocket ship is so long that light takes one second to traverse it from top to bottom. Finally, suppose there is an observer at the ceiling of the rocket ship and another at the floor, each with identical clocks that tick once each second.

Suppose the ceiling observer waits for the clock to tick, and then immediately sends a light signal down to the floor observer. The ceiling observer does this once more the next time the clock ticks. According to this setup, each signal travels for one second and then is received by the floor observer. So just as the ceiling observer sends two light signals a second apart, the floor observer receives two, one second apart.

How would this situation differ if the rocket ship were resting on earth, under the influence of gravity, instead of floating freely out in space? According to Newton's theory, gravity has no effect on this situation. If the observer on the ceiling sends signals one second apart, the observer will receive them one second apart. But the principle of equivalence does not make the same prediction. We can see what happens, that principle tells us, by considering the effect of uniform acceleration instead of the effect of gravity. This is an example of the way Einstein used the principle of equivalence to create his new theory of gravity.

So let's now suppose the rocket ship is accelerating. (We will imagine that it is accelerating slowly, so we don't approach the speed of light!) Since the rocket ship is moving upwards, the first signal will have less distance to travel than before and so will arrive sooner than one second later. If the rocket ship were moving at a constant speed, the second signal would arrive exactly the same amount of time sooner, so the time between the two signals would remain one second. But due to the acceleration, the rocket ship will be moving even faster when the second signal is sent than it was when the first signal was sent, so the second signal will have even less distance to traverse than the first and will arrive in even less time. The observer on the floor will therefore measure less than one second between the signals, disagreeing with the ceiling observer, who claims to have sent them exactly one second apart.

This is probably not startling in the case of the accelerating rocket ship—after all, we just explained it! But remember, the principle of equivalence says that it also applies to a rocket ship at rest in a gravitational field. That means that even if the rocket ship is not accelerating but, say, is sitting on a launching pad on the earth's surface, if the ceiling observer sends signals towards the floor at intervals of one each second (according to his clock), the floor observer will receive the signals at shorter intervals (according to his clock). That *is* startling!

You might still ask whether this means that gravity changes time, or whether it merely ruins clocks. Suppose the floor observer climbs up to the ceiling, where he and his partner compare their clocks. They are identical clocks, and sure enough, both observers will find that they now agree on the length of a second. There is nothing wrong with the floor observer's clock: it measures the local flow of time, wherever it happens to be. So just as special relativity tells us that time runs differently for observers in relative motion, general relativity tells us that time runs differently for observers at different heights in a gravitational field. According to general relativity, the floor observer measured less than one second between signals because time moves more slowly closer to the earth's surface. The stronger the field, the greater this effect. Newton's laws of motion put an end to the idea of absolute position in space. We have now seen how the theory of relativity gets rid of absolute time.

This prediction was tested in 1962, using a pair of very accurate clocks mounted at the top and bottom of a water tower. The clock at the bottom, which was nearer the earth, was found to run slower, in exact agreement with general relativity. The effect is a small one—a clock on the surface of the sun would gain only about a minute a year as compared to one on the surface of the earth. Yet with the advent of very accurate navigation systems based on signals from satellites, the difference in the speed of clocks at different heights above the earth is now of considerable practical importance. If you ignored the predictions of general relativity, the position that you calculated would be wrong by several miles!

Our biological clocks are equally affected by these changes in the flow of time. Consider a pair of twins. Suppose that one twin goes to live on the top of a mountain while the other stays at sea level. The first twin would age faster than the second. Thus, if they met again, one would be older than the other. In this case, the difference in ages would be very small, but it would be much larger if one of the twins went for a long trip in a spaceship in which he accelerated to nearly the speed of light. When he returned, he would be much younger than the one who stayed on earth. This is known as the twins paradox, but it is a paradox only if you have the idea of absolute time at the back of your mind. In the theory of relativity there is no unique absolute time; instead, each individual has his own personal measure of time that depends on where he is and how he is moving.

Before 1915, space and time were thought of as a fixed arena in which events took place but which was not affected by what happened in it. This was true even of the special theory of relativity. Bodies moved, forces attracted and repelled, but time and space simply continued unaffected. It was natural to think that space and time went on for ever. The situation, however, is quite different in the general theory of relativity. Space and time are now dynamic quantities: when a body moves or a force acts, it affects the curvature of space and time—and in turn the structure of space-time affects the way in which bodies move and forces act. Space and time not only affect but also are affected by everything that happens in the universe. Just as we cannot talk about events in the universe without the notions of space and time, so in general relativity it became meaningless to talk about space and time outside the limits of the universe. In the decades following 1915, this new understanding of space and time was to revolutionize our view of the universe. As we will see, the old idea of an essentially unchanging universe that could have existed for ever, and could continue to exist for ever, was replaced by the notion of a dynamic, expanding universe that seemed to have begun a finite time ago and which might end at a finite time in the future.

7

THE EXPANDING UNIVERSE

IF YOU LOOK AT THE SKY on a clear, moonless night, the brightest objects you see are likely to be the planets Venus, Mars, Jupiter, and Saturn. There will also be a very large number of stars, which are just like our own sun but much farther from us. Some of these fixed stars do, in fact, appear to change very slightly their positions relative to each other as the earth orbits around the sun. They are not really fixed at all! This is because they are comparatively near to us. As the earth goes around the sun, we see the nearer stars from different positions against the background of more distant stars. The effect is the same one you see when you are driving down an open road and the relative positions of nearby trees seem to change against the background of whatever is on the horizon. The nearer the trees, the more they seem to move. This change in relative position is called parallax. In the case of stars, it is fortunate, because it enables us to measure directly the distance of these stars from us.

As we mentioned in Chapter 1, the nearest star, Proxima Centauri, is about four light-years, or twenty-three million million miles, away. Most of the other stars that are visible to the naked eye lie within a few

Parallax

Whether you are moving down a road or through space, the relative position of nearer and farther objects changes as you go. A measure of that change can be used to determine the relative distance of the objects.

hundred light-years of us. Our sun, for comparison, is a mere eight light-minutes away! The visible stars appear spread all over the night sky but are particularly concentrated in one band, which we call the Milky Way. As long ago as 1750, some astronomers were suggesting that the appearance of the Milky Way could be explained if most of the visible stars lie in a single disklike configuration, one example of what we now call a spiral galaxy. Only a few decades later, the astronomer Sir William Herschel confirmed this idea by painstakingly cataloguing the positions and distances of vast numbers of stars. Even so, this idea gained complete acceptance only early in the twentieth century. We now know that the Milky Way—our galaxy—is about one hundred thousand light-years across and is slowly rotating; the stars in its spiral arms orbit around its centre about once every several hundred million years. Our sun is just an ordinary, average-sized yellow star near the inner edge of one of the spiral arms. We have certainly come a long way since Aristotle and Ptolemy, when we thought that the earth was the centre of the universe!

Our modern picture of the universe dates back only to 1924, when the American astronomer Edwin Hubble demonstrated that the Milky Way was not the only galaxy. He found, in fact, many others, with vast tracts of empty space between them. In order to prove this, Hubble needed to determine the distances from the earth to the other galaxies. But these galaxies were so far away that, unlike nearby stars, their positions really do appear fixed. Since Hubble couldn't use the parallax on these galaxies, he was forced to use indirect methods to measure their distances. One obvious measure of a star's distance is its brightness. But the apparent brightness of a star depends not only on its distance but also on how much light it radiates (its luminosity). A dim star, if near enough, will outshine the brightest star in any distant galaxy. So in order to use apparent brightness as a measure of its distance, we must know a star's luminosity.

The luminosity of nearby stars can be calculated from their apparent brightness because their parallax enables us to know their distance. Hubble noted that these nearby stars could be classified into certain types by the kind of light they give off. The same type of stars always had the same luminosity. He then argued that if we found these types of stars in a distant galaxy, we could assume that they had the same luminosity as the similar stars nearby. With that information, we could calculate the distance to that galaxy. If we could do this for a number of stars in the same galaxy and our calculations always gave the same distance, we could be fairly confident of our estimate. In this way, Hubble worked out the distances to nine different galaxies.

Today we know that stars visible to the naked eye make up only a minute fraction of all the stars. We can see about five thousand stars, only about .0001 per cent of all the stars in just our own galaxy, the Milky Way. The Milky Way itself is but one of more than a hundred billion galaxies that can be seen using modern telescopes—and each galaxy contains on average some one hundred billion stars. If a star were a grain of salt, you could fit all the stars visible to the naked eye on a teaspoon, but all the stars in the universe would fill a ball more than eight miles wide.

Stars are so far away that they appear to us to be just pinpoints of light. We cannot see their size or shape. But, as Hubble noticed, there are many different types of stars, and we can tell them apart by the colour of their light. Newton discovered that if light from the sun passes through a triangular piece of glass called a prism, it breaks up into its component colours as in a rainbow. The relative intensities of the various colours emitted by a given source of light are called its spectrum. By focusing a telescope on an individual star or galaxy, one can observe the spectrum of the light from that star or galaxy.

One thing this light tells us is temperature. In 1860, the German physicist Gustav Kirchhoff realized that any material body, such as a star, will give off light or other radiation when heated, just as coals glow when they are heated. The light such glowing objects give off is due to the thermal motion of the atoms within them. It is called black-body radiation (even though the glowing objects are not black).

Stellar Spectrum
By analysing the component colours of starlight, one can determine both the temperature of a star and the composition of its atmosphere.

The spectrum of black-body radiation is hard to mistake: it has a distinctive form that varies with the temperature of the body. The light emitted by a glowing object is therefore like a thermometer reading. The spectrum we observe from different stars is always in exactly this form: it is a postcard of the thermal state of that star.

If we look more closely, starlight tells us even more. We find that certain very specific colours are missing, and these missing colours may vary from star to star. Since we know that each chemical element absorbs a characteristic set of very specific colours, by matching these

Black-body Spectrum
All objects—not just stars—emit radiation resulting from the thermal motion of the objects' microscopic constituents. The distribution of frequencies in this radiation is characteristic of an object's temperature.

to those that are missing from a star's spectrum we can determine exactly which elements are present in that star's atmosphere.

In the 1920s, when astronomers began to look at the spectra of stars in other galaxies, they found something most peculiar: there were the same characteristic patterns of missing colours as for stars in our own galaxy, but they were all shifted towards the red end of the spectrum by the same relative amount.

To physicists, the shifting of colour or frequency is known as the

Doppler effect. We are all familiar with it in the realm of sound. Listen to a car passing on the road: as it approaches, its engine—or its horn—sounds at a higher pitch, and after it passes and is moving away, it sounds at a lower pitch. The sound of its engine or horn is a wave, a succession of crests and troughs. When a car is racing towards us, it will be progressively nearer to us as it emits each successive wave crest, so the distance between wave crests—the wavelength of the sound—will be smaller than if the car were stationary. The smaller the wavelength, the more of these fluctuations reach our ear each second, and the higher the pitch, or frequency, of the sound. Correspondingly, if the car is moving away from us, the wavelength will be greater and the waves will reach our ear with a lower frequency. The faster the car is moving, the greater the effect, so we can use the Doppler effect to measure speed. The behaviour of light or radio waves is similar. Indeed, the police make use of the Doppler effect to measure the speed of cars by measuring the wavelength of pulses of radio waves reflected off them.

As we noted in Chapter 5, the wavelength of visible light is extremely small, ranging from forty- to eighty-millionths of a centimetre. The different wavelengths of light are what the human eye sees as different colours, with the longest wavelengths appearing at the red end of the spectrum and the shortest wavelengths at the blue end. Now imagine a source of light at a constant distance from us, such as a star, emitting waves of light at a constant wavelength. The wavelength of the waves we receive will be the same as the wavelength at which they are emitted. Then suppose that the source starts to move away from us. As in the case of sound, this means that the light will have its wavelength elongated, and hence its spectrum will be shifted towards the red end of the spectrum.

In the years following his proof of the existence of other galaxies, Hubble spent his time cataloguing their distances and observing their spectra. At that time most people expected the galaxies to be moving around quite randomly, and so Hubble expected to find as many blue-

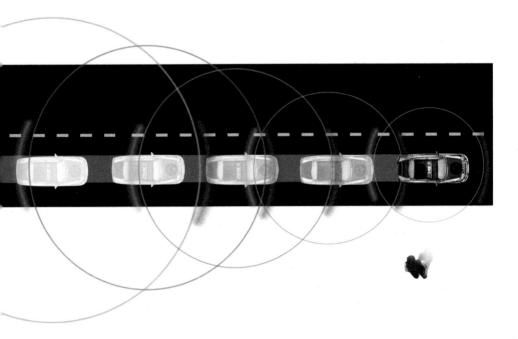

Doppler Effect
When a wave source moves towards an observer, its waves appear to have a shorter wavelength. If the wave source moves away, its waves appear to have a longer wavelength. This is called the Doppler effect.

shifted spectra as red-shifted ones. It was quite a surprise, therefore, to find that most galaxies appeared red-shifted: nearly all were moving away from us! More surprising still was the finding that Hubble published in 1929: even the size of a galaxy's red shift is not random but is directly proportional to the galaxy's distance from us. In other words, the farther a galaxy is, the faster it is moving away! And that meant that the universe could not be static or unchanging in size, as everyone previously had thought. It is in fact expanding; the distance between the different galaxies is growing all the time.

The discovery that the universe is expanding was one of the great intellectual revolutions of the twentieth century. With hindsight, it is easy to wonder why no one had thought of it before. Newton, and others, should have realized that a static universe would be unstable, for there is no comparable repulsive force to balance the gravitational pull that all the stars and galaxies exert upon each other. Therefore, even if at some time the universe had been static, it wouldn't have remained static because the mutual gravitational attraction of all the stars and galaxies would soon have started it contracting. In fact, even if the universe was expanding fairly slowly, the force of gravity would cause it eventually to stop expanding, and it would start to contract. However, if the universe was expanding faster than a certain critical rate, gravity would never be strong enough to stop it, and it would continue to expand for ever. This is a bit like what happens when you fire a rocket upwards from the surface of the earth. If the rocket has a fairly low speed, gravity will eventually stop it, and it will start falling back. On the other hand, if the rocket has more than a certain critical speed (about seven miles per second), gravity will not be strong enough to pull it back, so it will keep going away from the earth for ever.

This behaviour of the universe could have been predicted from Newton's theory of gravity at any time in the nineteenth, the eighteenth, or even the late seventeenth century. Yet so strong was the belief in a static universe that it persisted into the early twentieth century. Even Einstein, when he formulated the general theory of relativity in 1915, was so sure that the universe had to be static that he modified his theory to make this possible by introducing a fudge factor, called the cosmological constant, into his equations. The cosmological constant had the effect of a new "antigravity" force, which, unlike other forces, did not come from any particular source but was built into the very fabric of space-time. As a result of this new force, space-time had an inbuilt tendency to expand. By adjusting the cosmological constant, Einstein could adjust the strength of this tendency. He found he could adjust it to exactly balance the mutual attraction of all the matter in the universe, so a static universe would result. He later disavowed the cosmological constant, calling this fudge factor his "greatest mistake". As we'll soon see, today we have reason to believe that he might have been right to introduce it after all. But what must have disappointed Einstein was that he had allowed his belief in a static universe to override what his theory seemed to be predicting: that the universe is expanding. Only one man, it seems, was willing to take this prediction of general relativity at face value. While Einstein and other physicists were looking for ways of avoiding general relativity's non-static universe, the Russian physicist and mathematician Alexander Friedmann instead set about explaining it.

Friedmann made two very simple assumptions about the universe: that the universe looks identical in whichever direction we look, and that this would also be true if we were observing the universe from anywhere else. From these two ideas alone, Friedmann showed, by solving the equations of general relativity, that we should not expect the universe to be static. In fact, in 1922, several years before Edwin Hubble's discovery, Friedmann predicted exactly what Hubble later found!

The assumption that the universe looks the same in every direction is clearly not exactly true in reality. For example, as we have noted, the other stars in our galaxy form a distinct band of light across the night sky, called the Milky Way. But if we look at distant galaxies, there seems to be more or less the same number of them in every direction. So the universe does appear to be roughly the same in every direction, provided we view it on a large scale compared to the distance between galaxies, and ignore the differences on small scales. Imagine standing in a forest in which the trees are growing in random locations. If you look in one direction, you may see the nearest tree at a distance of one metre. In another direction, the nearest tree might be three metres away. In a third direction, you might see a clump of trees at two metres. It doesn't seem as if the forest looks the same in every direction, but if you were to take into account all the trees within a one-mile radius, these kinds of differences would average out and you would find that the forest is the same in whichever direction you look.

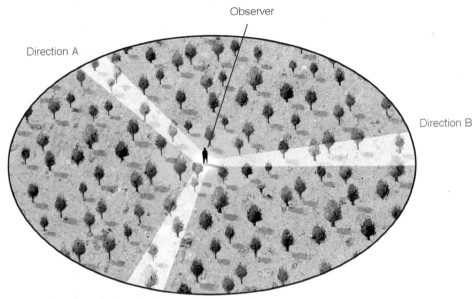

Isotropic Forest
Even if the trees in a forest are uniformly distributed, nearby trees may appear
bunched. Similarly, the universe does not look uniform in our local neighbourhood,
yet on large scales our view appears identical in whichever direction we look.

For a long time, the uniform distribution of stars was sufficient
justification for Friedmann's assumption—as a rough approximation
to the real universe. But more recently a lucky accident uncovered
another respect in which Friedmann's assumption is in fact a remark-
ably accurate description of our universe. In 1965, two American
physicists at the Bell Telephone Laboratories in New Jersey, Arno
Penzias and Robert Wilson, were testing a very sensitive microwave
detector. (Recall that microwaves are just like light waves, but with a
wavelength of around a centimetre.) Penzias and Wilson were worried when they found that their detector was picking up more noise
than it ought to. They discovered bird droppings in their detector and

checked for other possible malfunctions, but they soon ruled these out. The noise was peculiar in that it remained the same day and night and throughout the year, even though the earth was rotating on its axis and orbiting around the sun. Since the earth's rotation and orbit pointed the detector in different directions in space, Penzias and Wilson concluded that the noise was coming from beyond the solar system and even from beyond the galaxy. It seemed to be coming equally from every direction in space. We now know that in whichever direction we look, this noise never varies by more than a tiny fraction, so Penzias and Wilson had unwittingly stumbled across a striking example of Friedmann's first assumption that the universe is the same in every direction.

What is the origin of this cosmic background noise? At roughly the same time as Penzias and Wilson were investigating noise in their detector, two American physicists at nearby Princeton University, Bob Dicke and Jim Peebles, were also taking an interest in microwaves. They were working on a suggestion, made by George Gamow (once a student of Alexander Friedmann), that the early universe should have been very hot and dense, glowing white hot. Dicke and Peebles argued that we should still be able to see the glow of the early universe, because light from very distant parts of it would only just be reaching us now. However, the expansion of the universe meant that this light should be so greatly red-shifted that it would appear to us now as microwave radiation, rather than visible light. Dicke and Peebles were preparing to look for this radiation when Penzias and Wilson heard about their work and realized that they had already found it. For this, Penzias and Wilson were awarded the Nobel Prize in 1978 (which seems a bit hard on Dicke and Peebles, not to mention Gamow).

At first sight, all this evidence that the universe appears the same whichever direction we look in might seem to suggest there is something distinctive about our place in the universe. In particular, it might seem that if we observe all other galaxies to be moving away from us, then we must be at the centre of the universe. There is, however, an alternative explanation: the universe might look the same in every direction as seen from any other galaxy too. This, as we have seen, was Friedmann's second assumption.

We have no scientific evidence for or against that second assumption. Centuries ago, the church would have considered the assumption heresy, since church doctrine stated that we do occupy a special place at the centre of the universe. But today we believe Friedmann's assumption for almost the opposite reason, a kind of modesty: we feel it would be most remarkable if the universe looked the same in every direction around us but not around other points in the universe!

In Friedmann's model of the universe, all the galaxies are moving directly away from each other. The situation is rather like a balloon with a number of spots painted on it being steadily blown up. As the balloon expands, the distance between any two spots increases, but there is no spot that can be said to be the centre of the expansion. Moreover, as the radius of the balloon steadily increases, the farther apart the spots on the balloon, the faster they will be moving apart. For example, suppose that the radius of the balloon doubles in one second. Two spots that were previously one centimetre apart will now be two centimetres apart (as measured along the surface of the balloon), so their relative speed is one centimetre per second. On the other hand, a pair of spots that were separated by ten centimetres will now be separated by twenty, so their relative speed will be ten centimetres per second. Similarly, in Friedmann's model the speed at which any two galaxies are moving apart is proportional to the distance between them, so he predicted that the red shift of a galaxy should be directly proportional to its distance from us, exactly as Hubble found. Despite the success of his model and his prediction of Hubble's observations, Friedmann's work remained largely unknown in the West until similar models were discovered in 1935 by the American physicist Howard Robertson and the British mathematician Arthur Walker, in response to Hubble's discovery of the uniform expansion of the universe.

The Expanding Balloon Universe
As a result of the expansion of the universe, all galaxies are moving directly away from each other. Over time, like spots on an inflating balloon, galaxies that are farther apart increase their separation more than nearer galaxies. Hence, to an observer in any given galaxy, the more distant a galaxy is, the faster it appears to be moving.

Friedmann derived only one model of the universe. But if his assumptions are correct, there are actually three possible types of solutions to Einstein's equations, that is, three different kinds of Friedmann models—and three different ways the universe can behave.

In the first kind of solution (which Friedmann found), the universe is expanding sufficiently slowly that the gravitational attraction between the different galaxies causes the expansion to slow down and eventually to stop. The galaxies then start to move towards each other, and the universe contracts. In the second kind of solution, the universe is expanding so rapidly that the gravitational attraction can never stop it, though it does slow it down a bit. Finally, there is a third kind of solution, in which the universe is expanding only just fast enough to avoid collapse. The speed at which the galaxies are moving apart gets smaller and smaller, but it never quite reaches zero.

A remarkable feature of the first kind of Friedmann model is that in it the universe is not infinite in space, but neither does space have any boundary. Gravity is so strong that space is bent round on to itself. This is rather like the surface of the earth, which is finite but has no boundary. If you keep travelling in a certain direction on the surface of the earth, you never come up against an impassable barrier or fall over the edge, and you eventually come back to where you started. In this model, space is just like this, but with three dimensions instead of two for the earth's surface. The idea that you could go right round the universe and end up where you started makes good science fiction, but it doesn't have much practical significance, because it can be shown that the universe would collapse to zero size before you could get around. It is so large, you would need to travel faster than light in order to end up where you started before the universe came to an end—and that is not allowed! Space is also curved in the second Friedmann model, though in a different way. Only the third Friedmann model corresponds to a universe whose large-scale geometry is flat (though space is still curved, or warped, in the vicinity of massive objects).

Which Friedmann model describes our universe? Will the universe eventually stop expanding and start contracting, or will it expand for ever?

It turns out the answer to this question is more complicated than scientists first thought. The most basic analysis depends on two things: the present rate of expansion of the universe, and its present average density (the amount of matter in a given volume of space). The faster the current rate of expansion, the greater the gravitational force required to stop it, and thus the greater the density of matter needed. If the average density is greater than a certain critical value (determined by the rate of expansion), the gravitational attraction of the matter in the universe will succeed in halting its expansion and cause it to collapse—corresponding to the first Friedmann model. If the average density is less than the critical value, there is not enough gravitational pull to stop the expansion, and the universe will expand for ever—corresponding to Friedmann's second model. And if the average density of the universe is exactly the critical number, then the universe will for ever slow its expansion, ever more gradually approaching, but not ever reaching, a static size. This corresponds to the third Friedmann model.

So which is it? We can determine the present rate of expansion by measuring the velocities at which other galaxies are moving away from us, using the Doppler effect. This can be done very accurately. However, the distances to the galaxies are not very well known because we can measure them only indirectly. So all we know is that the universe is expanding by between 5 per cent and 10 per cent every billion years. Our uncertainty about the present average density of the universe is even greater. Still, if we add up the masses of all the stars that we can see in our galaxy and other galaxies, the total is less than one hundredth of the amount required to halt the expansion of the universe, even for the lowest estimate of the rate of expansion.

But that is not the whole story. Our galaxy and other galaxies must also contain a large amount of "dark matter" that we cannot see directly but which we know must be there because of the influence of its gravitational attraction on the orbits of stars in the galaxies. Perhaps the best evidence of this comes from the stars on the outskirts of spiral galaxies such as our Milky Way. These stars orbit their galaxies much too fast to be held in orbit merely by the gravitational attraction of the observed galactic stars. In addition, most galaxies are found in clusters, and we can similarly infer the presence of yet more dark matter in between the galaxies in these clusters by its effect on the motion of the galaxies. In fact, the amount of dark matter greatly exceeds the amount of ordinary matter in the universe. When we add up all this dark matter, we still get only about one-tenth of the amount of matter required to halt the expansion. But there could also be other forms of dark matter, distributed almost uniformly throughout the universe, that we have not yet detected and which might raise the average density of the universe even more. For instance, there exists a type of elementary particle called the neutrino, which interacts very weakly with matter and is extremely hard to detect (one recent neutrino experiment employed an underground detector filled with fifty thousand tons of water). The neutrino used to be thought massless, and therefore to have no gravitational attraction, but experiments over the last few years indicate that the neutrino actually does have a very tiny mass that had previously gone undetected. If neutrinos have mass, they could be a form of dark matter. Still, even allowing for neutrino dark matter, there appears to be far less matter in the universe than would be needed to halt its expansion, and so until recently most physicists would have agreed that the second type of Friedmann model applies.

Then came some new observations. In the last few years, several teams of researchers have studied tiny ripples in the background microwave radiation discovered by Penzias and Wilson. The size of those ripples can be used as an indicator of the large-scale geometry of the universe, and they appear to indicate that the universe is flat after all (as in the third Friedmann model)! Since there doesn't seem to be enough matter and dark matter to account for this, physicists have postulated the existence of another as yet undetected substance to explain it—dark energy.

To further complicate things, other recent observations indicate that the rate of expansion of the universe is actually not slowing down but speeding up. None of the Friedmann models does this! And it is very strange, since the effect of the matter in space, whether high or low density, can only be to slow the expansion. Gravity is, after all, attractive. For the cosmic expansion to be accelerating is something like the blast from a bomb gaining power rather than dissipating after the explosion. What force could be responsible for pushing the cosmos apart ever faster? No one is sure yet, but it could be evidence that

Einstein was right about the need for the cosmological constant (and its antigravity effects) after all.

With the rapid growth of new technologies and grand new satellite-borne telescopes, we are rapidly learning new and surprising things about the universe. We now have a good idea of its behaviour at late-time: the universe will continue to expand at an ever-increasing rate. Time will go on for ever, at least for those prudent enough not to fall into a black hole. But what about very early times? How did the universe begin, and what set it expanding?

8

THE BIG BANG, BLACK HOLES,
AND THE EVOLUTION
OF THE UNIVERSE

IN FRIEDMANN'S FIRST MODEL of the universe, the fourth dimension, time—like space—is finite in extent. It is like a line with two ends, or boundaries. So time has an end, and it also has a beginning. In fact, all solutions to Einstein's equations in which the universe has the amount of matter we observe share one very important feature: at some time in the past (about 13.7 billion years ago), the distance between neighbouring galaxies must have been zero. In other words, the entire universe was squashed into a single point with zero size, like a sphere of radius zero. At that time, the density of the universe and the curvature of space-time would have been infinite. It is the time that we call the big bang.

All our theories of cosmology are formulated on the assumption that space-time is smooth and nearly flat. That means that all our theories break down at the big bang: a space-time with infinite curvature can hardly be called nearly flat! Thus even if there were events before the big bang, we could not use them to determine what would happen afterward, because predictability would have broken down at the big bang.

Correspondingly, if, as is the case, we know only what has happened since the big bang, we cannot determine what happened beforehand. As far as we are concerned, events before the big bang can have no consequences and so should not form part of a scientific model of the universe. We should therefore cut them out of the model and say that the big bang was the beginning of time. This means that questions such as who set up the conditions for the big bang are not questions that science addresses.

Another infinity that arises if the universe has zero size is in temperature. At the big bang itself, the universe is thought to have been infinitely hot. As the universe expanded, the temperature of the radiation decreased. Since temperature is simply a measure of the average energy—or speed—of particles, this cooling of the universe would have a major effect on the matter in it. At very high temperatures, particles would be moving around so fast that they could escape any attraction towards each other resulting from nuclear or electromagnetic forces, but as they cooled off, we would expect particles that attract each other to start to clump together. Even the types of particles that exist in the universe depend on the temperature, and hence on the age, of the universe.

Aristotle did not believe that matter was made of particles. He believed that matter was continuous. That is, according to him, a piece of matter could be divided into smaller and smaller bits without any limit: there could never be a grain of matter that could not be divided further. A few Greeks, however, such as Democritus, held that matter was inherently grainy and that everything was made up of large numbers of various different kinds of atoms. (The word *atom* means "indivisible" in Greek.) We now know that this is true—at least in our environment, and in the present state of the universe. But the atoms of our universe did not always exist, they are not indivisible, and they represent only a small fraction of the types of particles in the universe.

Atoms are made of smaller particles: electrons, protons, and neutrons. The protons and neutrons themselves are made of yet smaller particles called quarks. In addition, corresponding to each of these subatomic particles there exists an antiparticle. Antiparticles have the same mass as their sibling particles but are opposite in their charge and other attributes. For instance, the antiparticle for an electron, called a positron, has a positive charge, the opposite of the charge of the electron. There could be whole antiworlds and antipeople made out of antiparticles. However, when an antiparticle and particle meet, they annihilate each other. So if you meet your antiself, don't shake hands—you would both vanish in a great flash of light!

Light energy comes in the form of another type of particle, a massless particle called a photon. The nearby nuclear furnace of the sun is the greatest source of photons for the earth. The sun is also a huge source of another kind of particle, the aforementioned neutrino (and antineutrino). But these extremely light particles hardly ever interact with matter, and hence they pass through us without effect, at a rate of billions each second. All told, physicists have discovered dozens of these elementary particles. Over time, as the universe has undergone a complex evolution, the makeup of this zoo of particles has also evolved. It is this evolution that has made it possible for planets such as the earth, and beings such as we, to exist.

One second after the big bang, the universe would have expanded enough to bring its temperature down to about ten billion degrees Celsius. This is about a thousand times the temperature at the centre of the sun, but temperatures as high as this are reached in H-bomb explosions. At this time the universe would have contained mostly photons, electrons, and neutrinos, and their antiparticles, together with some protons and neutrons. These particles would have had so much energy that when they collided, they would have produced many different particle/antiparticle pairs. For instance, colliding photons might produce an electron and its antiparticle, the positron. Some of these newly produced particles would collide with an antiparticle sibling and be annihilated. Any time an electron meets up with a positron, both will be annihilated, but the reverse process is not so easy: in order for two massless particles such as photons to create a particle/antiparticle pair such as an electron and a positron, the colliding massless particles must have a certain minimum energy. That is because an electron and positron have mass, and this newly created mass must come from the energy of the colliding particles. As the universe continued to expand and the temperature to drop, collisions having enough energy to create electron/positron pairs would occur less often than the rate at which the pairs were being destroyed by annihilation. So eventually most of the electrons and positrons would have annihilated each other to produce more photons, leaving only relatively few electrons.

Photon/Electron/Positron Equilibrium
In the early universe, there was a balance between pairs of electrons and positrons colliding to create photons, and the reverse process. As the temperature of the universe dropped, the balance was altered to favour photon creation. Eventually most electrons and positrons in the universe annihilated each other, leaving only the relatively few electrons present today.

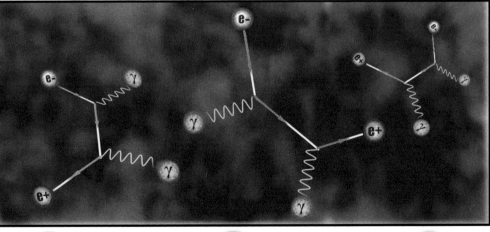

The neutrinos and antineutrinos, on the other hand, interact with themselves and with other particles only very weakly, so they would not annihilate each other nearly as quickly. They should still be around today. If we could observe them, it would provide a good test of this picture of a very hot early stage of the universe, but unfortunately, after billions of years their energies would now be too low for us to observe them directly (though we might be able to detect them indirectly).

About one hundred seconds after the big bang, the temperature of the universe would have fallen to one billion degrees, the temperature inside the hottest stars. At this temperature, a force called the strong force would have played an important role. The strong force, which we will discuss in more detail in Chapter 11, is a short-range attractive force that can cause protons and neutrons to bind to each other, forming nuclei. At high enough temperatures, protons and neutrons have enough energy of motion (see Chapter 5) that they can emerge from their collisions still free and independent. But at one billion degrees, they would no longer have had sufficient energy to overcome the attraction of the strong force, and they would have started to combine to produce the nuclei of atoms of deuterium (heavy hydrogen), which contain one proton and one neutron. The deuterium nuclei would then have combined with more protons and neutrons to make helium nuclei, which contain two protons and two neutrons, and also small amounts of a couple of heavier elements, lithium and beryllium. One can calculate that in the hot big bang model, about a quarter of the protons and neutrons would have been converted into helium nuclei, along with a small amount of heavy hydrogen and other elements. The remaining neutrons would have decayed into protons, which are the nuclei of ordinary hydrogen atoms.

This picture of a hot early stage of the universe was first put forward by the scientist George Gamow (see page 61) in a famous paper written in 1948 with a student of his, Ralph Alpher. Gamow had quite a sense of humour—he persuaded the nuclear scientist Hans Bethe to add his name to the paper to make the list of authors Alpher, Bethe, Gamow, like the first three letters of the Greek alphabet, alpha, beta, gamma, and particularly appropriate for a paper on the beginning of the universe! In this paper they made the remarkable prediction that radiation (in the form of photons) from the very hot early stages of the universe should still be around today, but with its temperature reduced to only a few degrees above absolute zero. (Absolute zero, −273 degrees Celsius, is the temperature at which substances contain no heat energy, and is thus the lowest possible temperature.)

It was this microwave radiation that Penzias and Wilson found in 1965. At the time that Alpher, Bethe, and Gamow wrote their paper, not much was known about the nuclear reactions of protons and neutrons. Predictions made for the proportions of various elements in the early universe were therefore rather inaccurate, but these calculations have been repeated in the light of better knowledge and now agree very well with what we observe. It is, moreover, very difficult to explain in any other way why about one-quarter of the mass of the universe is in the form of helium.

But there are problems with this picture. In the hot big bang model there was not enough time in the early universe for heat to have flowed from one region to another. This means that the initial state of the universe would have to have had exactly the same temperature everywhere in order to account for the fact that the microwave background has the same temperature in every direction we look. Moreover, the initial rate of expansion would have had to be chosen very precisely for the rate of expansion still to be so close to the critical rate needed to avoid collapse. It would be very difficult to explain why the universe should have begun in just this way, except as the act of a God who intended to create beings like us. In an attempt to find a model of the universe in which many different initial configurations could have evolved to something like the present universe, a scientist at the

Massachusetts Institute of Technology, Alan Guth, suggested that the early universe might have gone through a period of very rapid expansion. This expansion is said to be inflationary, meaning that the universe at one time expanded at an increasing rate. According to Guth, the radius of the universe increased by a million million million million million—1 with thirty zeros after it—times in only a tiny fraction of a second. Any irregularities in the universe would have been smoothed out by this expansion, just as the wrinkles in a balloon are smoothed away when you blow it up. In this way, inflation explains how the present smooth and uniform state of the universe could have evolved from many different non-uniform initial states. So we are therefore fairly confident that we have the right picture, at least going back to about one-billion-trillion-trillionth of a second after the big bang.

After all this initial turmoil, within only a few hours of the big bang, the production of helium and some other elements such as lithium would have stopped. And after that, for the next million years or so, the universe would have just continued expanding, without anything much happening. Eventually, once the temperature had dropped to a few thousand degrees and electrons and nuclei no longer had enough energy of motion to overcome the electromagnetic attraction between them, they would have started combining to form atoms. The universe as a whole would have continued expanding and cooling, but in regions that were slightly denser than average, this expansion would have been slowed down by the extra gravitational attraction.

This attraction would eventually stop expansion in some regions and cause them to start to collapse. As they were collapsing, the gravitational pull of matter outside these regions might start them rotating slightly. As the collapsing region got smaller, it would spin faster—just as skaters spinning on ice spin faster as they draw in their arms. Eventually, when the region got small enough, it would be spinning fast enough to balance the attraction of gravity, and in this way disklike rotating galaxies were born. Other regions that did not happen to pick up a rotation would become oval objects called elliptical galaxies. In these, the region would stop collapsing because individual parts of the galaxy would be orbiting stably around its centre, but the galaxy would have no overall rotation.

As time went on, the hydrogen and helium gas in the galaxies would break up into smaller clouds that would collapse under their own gravity. As these contracted and the atoms within them collided with one another, the temperature of the gas would increase, until eventually it became hot enough to start nuclear fusion reactions. These would convert the hydrogen into more helium. The heat released in this reaction, which is like a controlled hydrogen bomb explosion, is what makes a star shine. This additional heat also increases the pressure of the gas until it is sufficient to balance the gravitational attraction, and the gas stops contracting. In this manner, these clouds coalesce into stars like our sun, burning hydrogen into helium and radiating the resulting energy as heat and light. It is a bit like a balloon—there is a balance between the pressure of the air inside, which is trying to make the balloon expand, and the tension in the rubber, which is trying to make the balloon smaller.

Once clouds of hot gas coalesce into stars, the stars will remain stable for a long time, with heat from the nuclear reactions balancing the gravitational attraction. Eventually, however, the star will run out of its hydrogen and other nuclear fuels. Paradoxically, the more fuel a star starts off with, the sooner it runs out. This is because the more massive the star is, the hotter it needs to be to balance its gravitational attraction. And the hotter the star, the faster the nuclear fusion reaction and the sooner it will use up its fuel. Our sun has probably got enough fuel to last another five billion years or so, but more massive stars can use up their fuel in as little as one hundred million years, much less than the age of the universe.

When a star runs out of fuel, it starts to cool off and gravity takes over, causing it to contract. This contraction squeezes the atoms together and causes the star to become hotter again. As the star heats up further, it would start to convert helium into heavier elements such as carbon or oxygen. This, however, would not release much more energy, so a crisis would occur. What happens next is not completely clear, but it seems likely that the central regions of the star would collapse to a very dense state, such as a black hole. The term "black hole" is of very recent origin. It was coined in 1969 by the American scientist John Wheeler as a graphic description of an idea that goes back at least two hundred years, to a time when there were two theories about light: one, which Newton favoured, was that it was composed of particles, and the other was that it was made of waves. We now know that actually, both theories are correct. As we will see in Chapter 9, by the wave/particle duality of quantum mechanics, light can be regarded as both a wave and a particle. The descriptors *wave* and *particle* are concepts humans created, not necessarily concepts that nature is obliged to respect by making all phenomena fall into one category or the other!

Under the theory that light is made up of waves, it was not clear how it would respond to gravity. But if we think of light as being composed of particles, we might expect those particles to be affected by gravity in the same way that cannonballs, rockets, and planets are. In particular, if you shoot a cannonball upwards from the surface of the earth—or a star—like the rocket on page 58, it will eventually stop and then fall back unless the speed with which it starts upwards exceeds a certain value. This minimum speed is called the escape velocity. The escape velocity of a star depends on the strength of its gravitational pull. The more massive the star, the greater its escape velocity. At first people thought that particles of light travelled infinitely fast, so gravity would not have been able to slow them down, but the discovery by Roemer that light travels at a finite speed meant that gravity might have an important effect: if the star is massive enough, the speed of light will be less than the star's escape velocity, and all light emitted by the star will fall back into it. On this assumption, in 1783 a Cambridge don, John Michell, published a paper in the *Philosophical Transactions of*

Cannonballs Above and Below Escape Velocity
What goes up need not come down—if it is shot upwards faster than the escape velocity.

the Royal Society of London in which he pointed out that a star that was sufficiently massive and compact would have such a strong gravitational field that light could not escape: any light emitted from the surface of the star would be dragged back by the star's gravitational attraction before it could get very far. Such objects are what we now call black holes, because that is what they are: black voids in space.

A similar suggestion was made a few years later by a French scientist, the Marquis de Laplace, apparently independent of Michell. Interestingly, Laplace included it only in the first and second editions of his book *The System of the World,* leaving it out of later editions. Perhaps he decided it was a crazy idea—the particle theory of light went out of favour during the nineteenth century because it seemed that every-thing could be explained using the wave theory. In fact, it is not really consistent to treat light like cannonballs in Newton's theory of grav-ity because the speed of light is fixed. A cannonball fired upwards from the earth will be slowed down by gravity and will eventually stop and fall back; a photon, however, must continue upwards at a constant speed. A consistent theory of how gravity affects light did not come along until Einstein proposed general relativity in 1915, and the problem of understanding what would happen to a massive star, according to general relativity, was first solved by a young American, Robert Oppenheimer, in 1939.

The picture that we now have from Oppenheimer's work is as follows. The gravitational field of the star changes the paths of passing light rays in space-time from what they would have been had the star not been present. This is the effect that is seen in the bending of light from distant stars observed during an eclipse of the sun. The paths followed in space and time by light are bent slightly inwards near the surface of the star. As the star contracts, it becomes denser, so the gravitational field at its surface gets stronger. (You can think of the gravitational field as emanating from a point at the centre of the star; as the star shrinks, points on its surface get closer to the centre, so they feel a stronger field.) The stronger field makes light paths near the surface bend inwards more. Eventually, when the star has shrunk to a certain critical radius, the gravitational field at the surface becomes so strong that the light paths are bent inwards to the point that light can no longer escape.

According to the theory of relativity, nothing can travel faster than light. Thus if light cannot escape, neither can anything else; every-thing is dragged back by the gravitational field. The collapsed star has formed a region of space-time around it from which it is not possible to escape to reach a distant observer. This region is the black hole. The outer boundary of a black hole is called the event horizon. Today, thanks to the Hubble Space Telescope and other telescopes that focus on X-rays and gamma rays rather than visible light, we know that black holes are common phenomena—much more common than people first thought. One satellite discovered fifteen hundred black holes in just one small area of sky. We have also discovered a black hole in the centre of our galaxy, with a mass more than one million times that of our sun. That supermassive black hole has a star orbiting it at about 2 per cent the speed of light, faster than the average speed of an electron orbiting the nucleus in an atom!

In order to understand what you would see if you were watching a massive star collapse to form a black hole, it is necessary to remember that in the theory of relativity there is no absolute time. In other words, each observer has his own measure of time. The passage of time for someone on a star's surface will be different from that for someone at a distance, because the gravitational field is stronger on the star's surface.

Suppose an intrepid astronaut is on the surface of a collapsing star and stays on the surface as it collapses inwards. At some time on his watch—say, 11:00—the star would shrink below the critical radius at which the gravitational field becomes so strong that nothing can escape. Now suppose his instructions are to send a signal every second, according to his watch, to a spaceship above, which orbits at some fixed distance from the centre of the star. He begins transmitting at 10:59:58, that is, two seconds before 11:00. What will his companions on the spaceship record?

We learned from our earlier thought experiment aboard the rocket ship that gravity slows time, and the stronger the gravity, the greater the effect. The astronaut on the star is in a stronger gravitational field than his companions in orbit, so what to him is one second will be more than one second on their clocks. And as he rides the star's collapse inwards, the field he experiences will grow stronger and stronger, so the interval between his signals will appear successively longer to those on the spaceship. This stretching of time would be very small before 10:59:59, so the orbiting astronauts would have to wait only very slightly more than a second between the astronaut's 10:59:58 signal and the one that he sent when his watch read 10:59:59. But they would have to wait for ever for the 11:00 signal.

Everything that happens on the surface of the star between 10:59:59 and 11:00 (by the astronaut's watch) would be spread out over an infinite period of time, as seen from the spaceship. As 11:00 approached, the time interval between the arrival of successive crests and troughs of any light from the star would get successively longer, just as the interval between signals from the astronaut does. Since the frequency of light is a measure of the number of its crests and troughs per second, to those on the spaceship the frequency of the light from the star will get successively lower. Thus its light would appear redder and redder (and fainter and fainter). Eventually, the star would be so dim that it could no longer be seen from the spaceship: all that would be left would be a black hole in space. It would, however, continue to exert the same gravitational force on the spaceship, which would continue to orbit.

This scenario is not entirely realistic, however, because of the following problem. Gravity gets weaker the farther you are from the star, so the gravitational force on our intrepid astronaut's feet would always be greater than the force on his head. This difference in the forces would stretch him out like spaghetti or tear him apart before the star had contracted to the critical radius at which the event horizon formed! However, we believe that there are much larger objects in the universe, such as the central regions of galaxies, which can also undergo gravitational collapse to produce black holes, like the supermassive black hole at the centre of our galaxy. An astronaut on one of these would not be torn apart before the black hole formed. He would not, in fact, feel anything special as he reached the critical radius, and he could pass the point of no return without noticing it—though to those on the outside, his signals would again become further and further apart, and eventually stop. And within just a few hours (as measured by the astronaut), as the region continued to

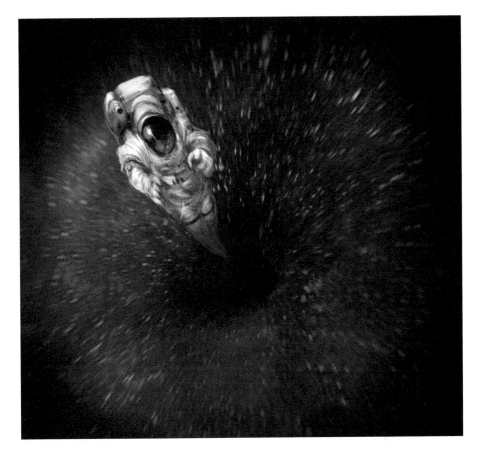

Tidal Forces
Since gravity weakens with distance, the earth pulls on your head with less force
than it pulls on your feet, which are a metre or two closer to the earth's centre.
The difference is so tiny we cannot feel it, but an astronaut near the surface of a black
hole would be literally torn apart.

collapse, the difference in the gravitational forces on his head and his
feet would become so strong that again it would tear him apart.

Sometimes, when a very massive star collapses, the outer regions
of the star may get blown off in a tremendous explosion called a supernova. A supernova explosion is so huge that it can give off more light
than all the other stars in its galaxy combined. One example of this is

the supernova whose remnants we see as the Crab Nebula. The Chinese recorded it in 1054. Though the star that exploded was five thousand light-years away, it was visible to the naked eye for months and shone so brightly that you could see it even during the day and read by it at night. A supernova five hundred light-years away—one-tenth as far—would be one hundred times brighter and could literally turn night into day. To understand the violence of such an explosion, just consider that its light would rival that of the sun, even though it is tens of millions of times farther away. (Recall that our sun resides at the neighbourly distance of eight light-minutes.) If a supernova were to occur close enough, it could leave the earth intact but still emit enough radiation to kill all living things. In fact, it was recently proposed that a die-off of marine creatures that occurred at the interface of the Pleistocene and Pliocene epochs about two million years ago was caused by cosmic ray radiation from a supernova in a nearby cluster of stars called the Scorpius–Centaurus association. Some scientists believe that advanced life is likely to evolve only in regions of galaxies in which there are not too many stars—"zones of life"— because in denser regions phenomena such as supernovas would be common enough to regularly snuff out any evolutionary beginnings. On the average, hundreds of thousands of supernovas explode somewhere in the universe each day. A supernova happens in any particular galaxy about once a century. But that's just the average. Unfortunately—for astronomers at least—the last supernova recorded in the Milky Way occurred in 1604, before the invention of the telescope.

The leading candidate for the next supernova explosion in our galaxy is a star called Rho Cassiopeiae. Fortunately, it is a safe and comfortable ten thousand light-years from us. It is in a class of stars known as yellow hypergiants, one of only seven known yellow hypergiants in the Milky Way. An international team of astronomers began to study this star in 1993. In the next few years they observed it undergoing periodic temperature fluctuations of a few hundred degrees.

Then in the summer of 2000, its temperature suddenly plummeted from around 7,000 degrees to 4,000 degrees Celsius. During that time, they also detected titanium oxide in the star's atmosphere, which they believe is part of an outer layer thrown off from the star by a massive shock wave.

In a supernova, some of the heavier elements produced near the end of the star's life are flung back into the galaxy and provide some of the raw material for the next generation of stars. Our own sun contains about 2 per cent of these heavier elements. It is a second- or third-generation star, formed some five billion years ago out of a cloud of rotating gas containing the debris of earlier supernovas. Most of the gas in that cloud went to form the sun or got blasted away, but small amounts of the heavier elements collected together to form the bodies that now orbit the sun as planets like the earth. The gold in our jewellry and the uranium in our nuclear reactors are both remnants of the supernovas that occurred before our solar system was born!

When the earth was newly condensed, it was very hot and without an atmosphere. In the course of time, it cooled and acquired an atmosphere from the emission of gases from the rocks. This early atmosphere was not one in which we could have survived. It contained no oxygen, but it did contain a lot of other gases that are poisonous to us, such as hydrogen sulphide (the gas that gives rotten eggs their smell). There are, however, other primitive forms of life that can flourish under such conditions. It is thought that they developed in the oceans, possibly as a result of chance combinations of atoms into large structures, called macromolecules, that were capable of assembling other atoms in the ocean into similar structures. They would thus have reproduced themselves and multiplied. In some cases there would be errors in the reproduction. Mostly these errors would have been such that the new macromolecule could not reproduce itself and eventually would have been destroyed. However, a few of the errors would have produced new macromolecules that were even better at reproducing themselves. They would have therefore had an advantage and would have tended to replace the original macro-molecules. In this way a process of evolution was started that led to the development of more and more complicated, self-reproducing organisms. The first primitive forms of life consumed various materials, including hydrogen sulphide, and released oxygen. This gradually changed the atmosphere to the composition that it has today, and allowed the development of higher forms of life such as fish, reptiles, mammals, and ultimately the human race.

The twentieth century saw man's view of the universe transformed: we realized the insignificance of our own planet in the vastness of the universe, and we discovered that time and space were curved and inseparable, that the universe was expanding, and that it had a beginning in time.

The picture of a universe that started off very hot and cooled as it expanded was based on Einstein's theory of gravity, general relativity. That it is in agreement with all the observational evidence that we have today is a great triumph for that theory. Yet because mathematics cannot really handle infinite numbers, by predicting that the universe began with the big bang, a time when the density of the universe and the curvature of space-time would have been infinite, the theory of general relativity predicts that there is a point in the universe where the theory itself breaks down, or fails. Such a point is an example of what mathematicians call a singularity. When a theory predicts singularities such as infinite density and curvature, it is a sign that the theory must somehow be modified. General relativity is an incomplete theory because it cannot tell us how the universe started off.

In addition to general relativity, the twentieth century also spawned another great partial theory of nature, quantum mechanics. That theory deals with phenomena that occur on very small scales. Our picture of the big bang tells us that there must have been a time in the very early universe when the universe was so small that, even when studying its large-scale structure, it was no longer possible to ignore the small-scale effects of quantum mechanics. We will see in the next chapter that our greatest hope for obtaining a complete understanding of the universe from beginning to end arises from combining these two partial theories into a single quantum theory of gravity, a theory in which the ordinary laws of science hold everywhere, including at the beginning of time, without the need for there to be any singularities.

9

QUANTUM GRAVITY

THE SUCCESS OF SCIENTIFIC THEORIES, particularly Newton's theory of gravity, led the Marquis de Laplace at the beginning of the nineteenth century to argue that the universe was completely deterministic. Laplace believed that there should be a set of scientific laws that would allow us—at least in principle—to predict everything that would happen in the universe. The only input these laws would need is the complete state of the universe at any one time. This is called an initial condition or a boundary condition. (A boundary can mean a boundary in space or time; a boundary condition in space is the state of the universe at its outer boundary—if it has one.) Based on a complete set of laws and the appropriate initial or boundary condition, Laplace believed, we should be able to calculate the complete state of the universe at any time.

The requirement of initial conditions is probably intuitively obvious: different states of being at present will obviously lead to different future states. The need for boundary conditions in space is a little more subtle, but the principle is the same. The equations on which physical theories are based can generally have very different solutions, and you must rely on the initial or boundary conditions to decide which solutions apply. It's a little like saying that your bank account has large amounts going in and out of it. Whether you end up bankrupt or rich depends not only on the sums paid in and out but also on the boundary or initial condition of how much was in the account to start with.

If Laplace were right, then, given the state of the universe at the present, these laws would tell us the state of the universe in both the future and the past. For example, given the positions and speeds of the sun and the planets, we can use Newton's laws to calculate the state of the solar system at any later or earlier time. Determinism seems fairly obvious in the case of the planets—after all, astronomers are very accurate in their predictions of events such as eclipses. But Laplace went further to assume that there were similar laws governing everything else, including human behaviour.

Is it really possible for scientists to calculate what all our actions will be in the future? A glass of water contains more than 10^{24} molecules (a 1 followed by twenty-four zeros). In practice we can never hope to know the state of each of these molecules, much less the complete state of the universe or even of our bodies. Yet to say that the universe is deterministic means that even if we don't have the brainpower to do the calculation, our futures are nevertheless predetermined.

This doctrine of scientific determinism was strongly resisted by many people, who felt that it infringed God's freedom to make the world run as He saw fit. But it remained the standard assumption of science until the early years of the twentieth century. One of the first indications that this belief would have to be abandoned came when the British scientists Lord Rayleigh and Sir James Jeans calculated the amount of black-body radiation that a hot object such as a star must radiate. (As noted in Chapter 7, any material body, when heated, will give off black-body radiation.)

According to the laws we believed at the time, a hot body ought to give off electromagnetic waves equally at all frequencies. If this were true, then it would radiate an equal amount of energy in every colour of the spectrum of visible light, and for all frequencies of microwaves, radio waves, X-rays, and so on. Recall that the frequency of a wave is the number of times per second that the wave oscillates up and down, that is, the number of waves per second. Mathematically, for a hot body to give off waves equally at all frequencies means that a hot body should radiate the same amount of energy in waves with frequencies between zero and one million waves per second as it does in waves with frequencies between one million and two million waves per second, two million and three million waves per second, and so forth, going on for ever. Let's say that one unit of energy is radiated in waves with frequencies between zero and one million waves per second, and in waves with frequencies between one million and two million waves per second, and so on. The total amount of energy radiated in all frequencies would then be the sum 1 plus 1 plus 1 plus . . . going on for ever. Since the number of waves per second in a wave is unlimited, the sum of energies is an unending sum. According to this reasoning, the total energy radiated should be infinite.

In order to avoid this obviously ridiculous result, the German scientist Max Planck suggested in 1900 that light, X-rays, and other electromagnetic waves could be given off only in certain discrete packets, which he called quanta. Today, as mentioned in Chapter 8, we call a quantum of light a photon. The higher the frequency of light, the greater its energy content. Therefore, though photons of any given colour or frequency are all identical, Planck's theory states that photons of different frequencies are different in that they carry different amounts of energy. This means that in quantum theory the faintest light of any given colour—the light carried by a single photon—has an energy content that depends upon its colour. For example, since violet light has twice the frequency of red light, one quantum of violet light has twice the energy content of one quantum of red light. Thus the smallest possible bit of violet light energy is twice as large as the smallest possible bit of red light energy.

How does this solve the black-body problem? The smallest amount of electromagnetic energy a black body can emit in any given frequency is that carried by one photon of that frequency. The energy of a photon is greater at higher frequencies. Thus the smallest amount of energy a black body can emit is higher at higher frequencies. At high enough frequencies, the amount of energy in even a single quantum will be more than a body has available, in which case no light will be emitted, ending the previously unending sum. Thus in Planck's theory, the radiation at high frequencies would be reduced, so the rate at which the body lost energy would be finite, solving the black-body problem.

The quantum hypothesis explained the observed rate of emission of radiation from hot bodies very well, but its implications for determinism were not realized until 1926, when another German scientist, Werner Heisenberg, formulated his famous uncertainty principle.

Faintest Possible Light

Faint light means fewer photons. The faintest possible light of any colour is the light carried by a single photon.

The uncertainty principle tells us that, contrary to Laplace's belief, nature does impose limits on our ability to predict the future using scientific law. This is because, in order to predict the future position and velocity of a particle, one has to be able to measure its initial state—that is, its present position and its velocity—accurately. The obvious way to do this is to shine light on the particle. Some of the waves of light will be scattered by the particle. These can be detected by the observer and will indicate the particle's position. However, light of a given wavelength has only limited sensitivity: you will not be able to determine the position of the particle more accurately than the distance between the wave crests of the light. Thus, in order to measure the position of the particle precisely, it is necessary to use light of a short wavelength, that is, of a high frequency. By Planck's quantum hypothesis, though, you cannot use an arbitrarily small amount of light: you have to use at least one quantum, whose energy is higher at higher frequencies. Thus, the more accurately you wish to measure the position of a particle, the more energetic the quantum of light you must shoot at it.

According to quantum theory, even one quantum of light will disturb the particle: it will change its velocity in a way that cannot be predicted. And the more energetic the quantum of light you use, the greater the likely disturbance. That means that for more precise measurements of position, when you will have to employ a more energetic quantum, the velocity of the particle will be disturbed by a larger amount. So the more accurately you try to measure the position of the particle, the less accurately you can measure its speed, and vice versa. Heisenberg showed that the uncertainty in the position of the particle times the uncertainty in its velocity times the mass of the particle can never be smaller than a certain fixed quantity. That means, for instance, if you halve the uncertainty in position, you must double the uncertainty in velocity, and vice versa. Nature forever constrains us to making this trade-off.

How bad is this trade-off? That depends on the numerical value of the "certain fixed quantity" we mentioned above. That quantity is known as Planck's constant, and it is a very tiny number. Because

Planck's constant is so tiny, the effects of the trade-off, and of quantum theory in general, are, like the effects of relativity, not directly noticeable in our everyday lives. (Though quantum theory does affect our lives—as the basis of such fields as, say, modern electronics.) For example, if we pinpoint the position of a Ping-Pong ball with a mass of one gram to within one centimetre in any direction, then we can pinpoint its speed to an accuracy far greater than we would ever need to know. But if we measure the position of an electron to an accuracy of roughly the confines of an atom, then we cannot know its speed more precisely than about plus or minus one thousand kilometres per second, which is not very precise at all.

The limit dictated by the uncertainty principle does not depend on the way in which you try to measure the position or velocity of the particle, or on the type of particle. Heisenberg's uncertainty principle is a fundamental, inescapable property of the world, and it has had profound implications for the way in which we view the world. Even after more than seventy years, these implications have not been fully appreciated by many philosophers and are still the subject of much controversy. The uncertainty principle signalled an end to Laplace's dream of a theory of science, a model of the universe that would be completely deterministic. We certainly cannot predict future events exactly if we cannot even measure the present state of the universe precisely!

We could still imagine that there is a set of laws that determine events completely for some supernatural being who, unlike us, could observe the present state of the universe without disturbing it. However, such models of the universe are not of much interest to us ordinary mortals. It seems better to employ the principle of economy known as Occam's razor and cut out all the features of the theory that cannot be observed. This approach led Heisenberg, Erwin Schrödinger, and Paul Dirac in the 1920s to reformulate Newton's mechanics into a new theory called quantum mechanics, based on the uncertainty principle. In this theory, particles no longer had separate, well-defined positions and velocities. Instead, they had a quantum state, which was a combination of position and velocity defined only within the limits of the uncertainty principle.

One of the revolutionary properties of quantum mechanics is that it does not predict a single definite result for an observation. Instead, it predicts a number of different possible outcomes and tells us how likely each of these is. That is to say, if you made the same measurement on a large number of similar systems, each of which started off in the same way, you would find that the result of the measurement would be A in a certain number of cases, B in a different number, and so on. You could predict the approximate number of times that the result would be A or B, but you could not predict the specific result of an individual measurement.

For instance, imagine you toss a dart towards a dartboard. According to classical theories—that is, the old, non-quantum theories—the dart will either hit the bull's-eye or it will miss it. And if you know the velocity of the dart when you toss it, the pull of gravity, and other such factors, you'll be able to calculate whether it will hit or miss. But quantum theory tells us this is wrong, that you cannot say it for certain. Instead, according to quantum theory there is a certain probability that the dart will hit the bull's-eye, and also a non-zero probability that it will land in any other given area of the board. Given an object as large as a dart, if the classical theory—in this case Newton's laws—says the dart will hit the bull's-eye, then you can be safe in assuming it will. At least, the chances that it won't (according to quantum theory) are so small that if you went on tossing the dart in exactly the same manner until the end of the universe, it is probable that you would still never observe the dart missing its target. But on the atomic scale, matters are different. A dart made of a single atom might have a 90 per cent probability of hitting the bull's-eye, with a 5 per cent chance of hitting elsewhere on the board, and another 5 per cent chance of missing it completely. You cannot say in advance which of these it will be. All you can say is that if you repeat the experiment many times, you can expect that, on average, ninety times out of each hundred times you repeat the experiment, the dart will hit the bull's-eye.

Quantum mechanics therefore introduces an unavoidable element of unpredictability or randomness into science. Einstein objected to this very strongly, despite the important role he had played in the development of these ideas. In fact, he was awarded the Nobel Prize for his contribution to quantum theory. Nevertheless, he never accepted that the universe was governed by chance; his feelings were summed up in his famous statement "God does not play dice."

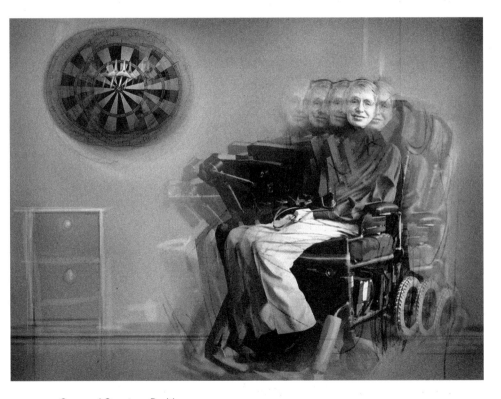

Smeared Quantum Position
According to quantum theory, one cannot pinpoint an object's position and velocity with infinite precision, nor can one predict exactly the course of future events.

The test of a scientific theory, as we have said, is its ability to predict the results of an experiment. Quantum theory limits our abilities. Does this mean quantum theory limits science? If science is to progress, the way we carry it on must be dictated by nature. In this case, nature requires that we redefine what we mean by prediction: We may not be able to predict the outcome of an experiment exactly, but we can repeat the experiment many times and confirm that the various possible outcomes occur within the probabilities predicted by quantum theory. Despite the uncertainty principle, therefore, there is no need to give up on the belief in a world governed by physical law. In fact, in the end, most scientists were willing to accept quantum mechanics precisely because it agreed perfectly with experiment.

One of the most important implications of Heisenberg's uncertainty principle is that particles behave in some respects like waves. As we have seen, they do not have a definite position but are "smeared out" with a certain probability distribution. Equally, although light is made up of waves, Planck's quantum hypothesis also tells us that in some ways light behaves as if it were composed of particles: it can be emitted or absorbed only in packets, or quanta. In fact, the theory of quantum mechanics is based on an entirely new type of mathematics that no longer describes the real world in terms of either particles or waves. For some purposes it is helpful to think of particles as waves and for other purposes it is better to think of waves as particles, but these ways of thinking are just conveniences. This is what physicists mean when they say there is a duality between waves and particles in quantum mechanics.

An important consequence of wavelike behaviour in quantum mechanics is that one can observe what is called interference between two sets of particles. Normally, interference is thought of as a phenomenon of waves; that is to say, when waves collide, the crests of one set of waves may coincide with the troughs of the other set, in which case the waves are said to be out of phase. If that happens, the two sets of waves then cancel each other out, rather than adding up to a stronger wave, as one might expect. A familiar example of

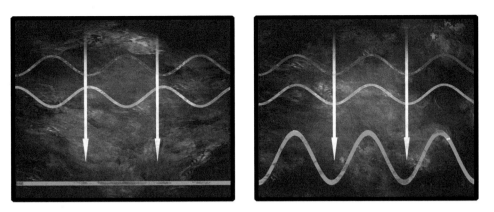

In and Out of Phase
If the crests and troughs of two waves coincide, they result in a stronger wave, but if
one wave's crests coincide with another's troughs, the two waves cancel each other.

interference in the case of light is the colours that are often seen in soap bubbles. These are caused by reflection of light from the two sides of the thin film of water forming the bubble. White light consists of light waves of all different wavelengths, or colours. For certain wavelengths the crests of the waves reflected from one side of the soap film coincide with the troughs reflected from the other side. The colours corresponding to these wavelengths are absent from the reflected light, which therefore appears to be coloured.

But quantum theory tells us that interference can also occur for particles, because of the duality introduced by quantum mechanics. A famous example is the so-called two-slit experiment. Imagine a partition—a thin wall—with two narrow parallel slits in it. Before we consider what happens when particles are sent through these slits, let's examine what happens when light is shined on them. On one side of the partition you place a source of light of a particular colour (that is, of a particular wavelength). Most of the light will hit the partition, but a small amount will go through the slits. Now suppose you place a screen on the far side of the partition from the light. Any point on that screen will receive waves from both slits. However, in general, the distance the light has to travel from the light source to the point via one of the slits will be different than for the light travelling via the other slit. Since the distance travelled differs, the waves from the two slits will not be in phase with each other when they arrive at the point. In some places the troughs from one wave will coincide with the crests from the other, and the waves will cancel each other out; in other places the crests and troughs will coincide, and the waves will reinforce each other; and in most places the situation will be somewhere in between. The result is a characteristic pattern of light and dark.

The remarkable thing is that you get exactly the same kind of pattern if you replace the source of light by a source of particles, such as electrons, that have a definite speed. (According to quantum theory, if the electrons have a definite speed the corresponding matter waves

Path Distances and Interference
In the two-slit experiment, the distance that waves must travel from the top and bottom slits to the screen varies with height along the screen. The result is that the waves reinforce each other at certain heights and cancel at others, forming an interference pattern.

have a definite wavelength.) Suppose you have only one slit and start firing electrons at the partition. Most of the electrons will be stopped by the partition, but some will go through the slit and make it to the screen on the other side. It might seem logical to assume that opening a second slit in the partition would simply increase the number of electrons hitting each point of the screen. But if you open the second slit, the number of electrons hitting the screen increases at some points and decreases at others, just as if the electrons were interfering as waves do, rather than acting as particles.

Now imagine sending the electrons through the slits one at a time. Is there still interference? One might expect each electron to pass through one slit or the other, doing away with the interference pattern. In reality, however, even when the electrons are sent through one at a time, the interference pattern still appears. Each electron, therefore, must be passing through both slits at the same time and interfering with itself!

The phenomenon of interference between particles has been crucial to our understanding of the structure of atoms, the basic units out of which we, and everything around us, are made. In the early twentieth century it was thought that atoms were rather like the planets orbiting the sun, with electrons (particles of negative electricity) orbiting around a central nucleus, which carried positive electricity. The attraction between the positive and negative electricity was supposed to keep the electrons in their orbits in the same way that the gravitational attraction between the sun and the planets keeps the planets in their orbits. The trouble with this was that the classical laws of mechanics and electricity, before quantum mechanics, predicted that electrons orbiting in this manner would give off radiation. This would cause them to lose energy and thus spiral inward until they collided with the nucleus. This would mean that the atom, and indeed all matter, should rapidly collapse to a state of very high density, which obviously doesn't happen!

The Danish scientist Niels Bohr found a partial solution to this problem in 1913. He suggested that perhaps the electrons were not

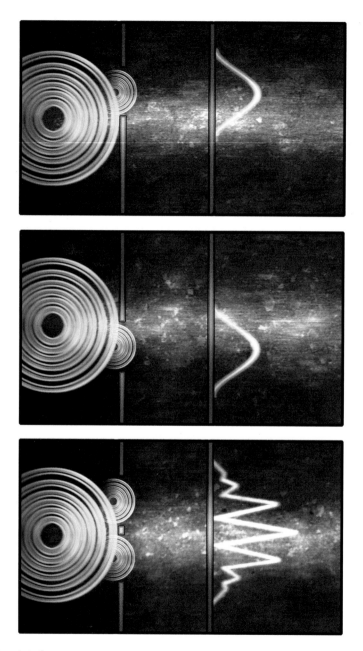

Electron Interference

Because of interference, the result of sending a beam of electrons through two slits does not correspond to the result of sending the electrons through each slit separately.

able to orbit at just any distance from the central nucleus but rather could orbit only at certain specified distances. Supposing that only one or two electrons could orbit at any one of these specified distances would solve the problem of the collapse, because once the limited number of inner orbits was full, the electrons could not spiral in any farther. This model explained quite well the structure of the simplest atom, hydrogen, which has only one electron orbiting around the nucleus. But it was not clear how to extend this model to more complicated atoms. Moreover, the idea of a limited set of allowed orbits seemed like a mere Band-Aid. It was a trick that worked mathematically, but no one knew why nature should behave that way, or what deeper law—if any—it represented. The new theory of quantum mechanics resolved this difficulty. It revealed that an electron orbiting around the nucleus could be thought of as a wave, with a wavelength that depended on its velocity. Imagine the wave circling the nucleus at specified distances, as Bohr had postulated. For certain orbits, the circumference of the orbit would correspond to a whole number (as opposed to a fractional number) of wavelengths of the electron. For these orbits the wave crest would be in the same position each time round, so the waves would reinforce each other. These orbits would correspond to Bohr's allowed orbits. However, for orbits whose lengths were not a whole number of wavelengths, each wave crest would eventually be cancelled out by a trough as the electrons went round. These orbits would not be allowed. Bohr's law of allowed and forbidden orbits now had an explanation.

A nice way of visualizing the wave/particle duality is the so-called sum over histories introduced by the American scientist Richard Feynman. In this approach a particle is not supposed to have a single history or path in space-time, as it would in a classical, non-quantum theory. Instead it is supposed to go from point A to point B by every possible path. With each path between A and B, Feynman associated a couple of numbers. One represents the amplitude, or size, of a wave. The other represents the phase, or position in the cycle (that is, whether

Waves in Atomic Orbits
Niels Bohr imagined the atom as consisting of electron waves endlessly circling atomic nuclei. In his picture, only orbits with circumferences corresponding to a whole number of electron wavelengths could survive without destructive interference.

it is at a crest or a trough or somewhere in between). The probability of a particle going from A to B is found by adding up the waves for all the paths connecting A and B. In general, if one compares a set of neighbouring paths, the phases or positions in the cycle will differ greatly. This means that the waves associated with these paths will almost exactly cancel each other out. However, for some sets of neighbouring paths the phase will not vary much between paths, and the waves for these paths will not cancel out. Such paths correspond to Bohr's allowed orbits.

With these ideas in concrete mathematical form, it was relatively straightforward to calculate the allowed orbits in more complicated atoms and even in molecules, which are made up of a number of atoms held together by electrons in orbits that go around more than one nucleus. Since the structure of molecules and their reactions with each other underlie all of chemistry and biology, quantum mechanics allows us in principle to predict nearly everything we see around us, within the limits set by the uncertainty principle. (In practice, however, we

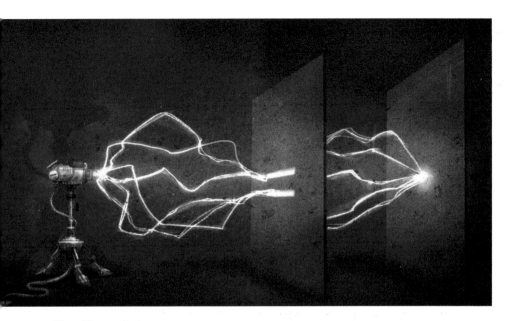

Many Electron Paths
In Richard Feynman's formulation of quantum theory, a particle, such as this one
moving from source to screen, takes every possible path.

cannot solve the equations for any atom besides the simplest one, hydrogen, which has only one electron, and we use approximations and
computers to analyse more complicated atoms and molecules.)

Quantum theory has been an outstandingly successful theory and
underlies nearly all of modern science and technology. It governs the
behaviour of transistors and integrated circuits, which are the essential
components of electronic devices such as televisions and computers,
and it is also the basis of modern chemistry and biology. The only
areas of physical science into which quantum mechanics has not yet
been properly incorporated are gravity and the large-scale structure
of the universe: Einstein's general theory of relativity, as noted earlier,
does not take account of the uncertainty principle of quantum
mechanics, as it should for consistency with other theories.

As we saw in the last chapter, we already know that general relativity must be altered. By predicting points of infinite density—singularities—classical (that is, non-quantum) general relativity predicts its own downfall, just as classical mechanics predicted its downfall by suggesting that black bodies should radiate infinite energy or that atoms should collapse to infinite density. And as with classical mechanics, we hope to eliminate these unacceptable singularities by making classical general relativity into a quantum theory—that is, by creating a quantum theory of gravity.

If general relativity is wrong, why have all experiments thus far supported it? The reason that we haven't yet noticed any discrepancy with observation is that all the gravitational fields that we normally experience are very weak. But as we have seen, the gravitational field should get very strong when all the matter and energy in the universe are squeezed into a small volume in the early universe. In the presence of such strong fields, the effects of quantum theory should be important.

Although we do not yet possess a quantum theory of gravity, we do know a number of features we believe it should have. One is that it should incorporate Feynman's proposal to formulate quantum theory in terms of a sum over histories. A second feature that we believe must be part of any ultimate theory is Einstein's idea that the gravitational field is represented by curved space-time: particles try to follow the nearest thing to a straight path in a curved space, but because space-time is not flat, their paths appear to be bent, as if by a gravitational field. When we apply Feynman's sum over histories to Einstein's view of gravity, the analogue of the history of a particle is now a complete curved space-time that represents the history of the whole universe.

In the classical theory of gravity, there are only two possible ways the universe can behave: either it has existed for an infinite time, or else it had a beginning at a singularity at some finite time in the past. For reasons we discussed earlier, we believe that the universe has not existed for ever. Yet if it had a beginning, according to classical general relativity, in order to know which solution of Einstein's equations describes our universe, we must know its initial state—that is, exactly how the universe began. God may have originally decreed the laws of nature, but it appears that He has since left the universe to evolve according to them and does not now intervene in it. How did He choose the initial state or configuration of the universe? What were the boundary conditions at the beginning of time? In classical general relativity this is a problem, because classical general relativity breaks down at the beginning of the universe.

In the quantum theory of gravity, on the other hand, a new possibility arises that, if true, would remedy this problem. In the quantum theory, it is possible for space-time to be finite in extent and yet to have no singularities that formed a boundary or edge. Space-time would be like the surface of the earth, only with two more dimensions. As was pointed out before, if you keep travelling in a certain direction on the surface of the earth, you never come up against an impassable barrier or fall over the edge, but eventually come back to where you started, without running into a singularity. So if this turns out to be the case, then the quantum theory of gravity has opened up a new possibility in which there would be no singularities at which the laws of science broke down.

If there is no boundary to space-time, there is no need to specify the behaviour at the boundary—no need to know the initial state of the universe. There is no edge of space-time at which we would have to appeal to God or some new law to set the boundary conditions for space-time. We could say: "The boundary condition of the universe is that it has no boundary." The universe would be completely self-contained and not affected by anything outside itself. It would be neither created nor destroyed. It would just BE. As long as we believed the universe had a beginning, the role of a creator seemed clear. But if the universe is really completely self-contained, having no boundary or edge, having neither beginning nor end, then the answer is not so obvious: what is the role of a creator?

10

WORMHOLES AND
TIME TRAVEL

IN PREVIOUS CHAPTERS WE HAVE SEEN how our views of the nature of time have changed over the years. Until the beginning of the twentieth century, people believed in an absolute time. That is, each event could be labelled by a number called "time" in a unique way, and all good clocks would agree on the time interval between two events. However, the discovery that the speed of light appeared the same to every observer, no matter how he was moving, led to the theory of relativity—and the abandoning of the idea that there was a unique absolute time. The time of events could not be labelled in a unique way. Instead, each observer would have his own measure of time as recorded by a clock that he carried, and clocks carried by different observers would not necessarily agree. Thus time became a more personal concept, relative to the observer who measured it. Still, time was treated as if it were a straight railway line on which you could go only one way or the other. But what if the railway line had loops and branches, so a train could keep going forwards but come back to a station it had already passed? In other words, might it be possible for someone to travel into the future or the past? H. G. Wells in *The Time Machine* explored these possibilities, as have countless other writers of

Time Machine
The authors in a time machine.

science fiction. Yet many of the ideas of science fiction, like submarines and travel to the moon, have become matters of science fact. So what are the prospects for time travel?

It is possible to travel to the future. That is, relativity shows that it is possible to create a time machine that will jump you forward in time. You step into the time machine, wait, step out, and find that much more time has passed on the earth than has passed for you. We do not have the technology today to do this, but it is just a matter of engineering: we know it can be done. One method of building such a machine would be to exploit the situation we discussed in Chapter 6 regarding the twins paradox. In this method, while you are sitting in the time machine, it

would be at least a hundred thousand years before it came back. Not a good situation if you want to write about intergalactic warfare! Still, the theory of relativity does allow one consolation, again along the lines of our discussion of the twins paradox in Chapter 6: it is possible for the journey to seem to be much shorter for the space travellers than for those who remain on earth. But there would not be much joy in returning from a space voyage a few years older to find that everyone you had left behind was dead and gone thousands of years ago. So in order to have any human interest in their stories, science fiction writers had to suppose that we would one day discover how to travel faster than light. Most of these authors don't seem to have realized the fact that if you can travel faster than light, the theory of relativity implies you can also travel back in time, as the following limerick says:

> There was a young lady of Wight
> Who travelled much faster than light.
> She departed one day,
> In a relative way,
> And arrived on the previous night.

The key to this connection is that the theory of relativity says not only that there is no unique measure of time on which all observers will agree but that, under certain circumstances, observers need not even agree on the order of events. In particular, if two events, A and B, are so far away in space that a rocket must travel faster than the speed of light to get from event A to event B, then two observers moving at different speeds can disagree on whether event A occurred before B, or event B occurred before event A. Suppose, for instance, that event A is the finish of the final hundred-metre race of the Olympic Games in 2012 and event B is the opening of the 100,004th meeting of the Congress of Proxima Centauri. Suppose that to an observer on Earth, event A happened first, and then event B. Let's say that B happened a year later, in 2013 by Earth's time. Since the earth and Proxima Centauri are some four light-years apart, these two events satisfy the above criterion: though A happens before B, to get from A to B you would have to travel faster than light. Then, to an observer on Proxima Centauri moving away from the earth at nearly the speed of light, it would appear that the order of the events is reversed: it appears that event B occurred before event A. This observer would say it is possible, if you could move faster than light, to get from event B to event A. In fact, if you went really fast, you could also get back from A to Proxima Centauri before the race and place a bet on it in the sure knowledge of who would win!

There is a problem with breaking the speed-of-light barrier. The theory of relativity says that the rocket power needed to accelerate a spaceship gets greater and greater the nearer it gets to the speed of light. We have experimental evidence for this, not with spaceships but with elementary particles in particle accelerators such as those at Fermilab or the European Centre for Nuclear Research (CERN). We can accelerate particles to 99.99 per cent of the speed of light, but however much power we feed in, we can't get them beyond the speed-of-light barrier. Similarly with spaceships: no matter how much rocket power they have, they can't accelerate beyond the speed of light. And since travel backwards in time is possible only if faster-than-light travel is possible, this might seem to rule out both rapid space travel and travel back in time.

However, there is a possible way out. It might be that you could warp space-time so that there was a shortcut between A and B. One way of doing this would be to create a wormhole between A and B. As its name suggests, a wormhole is a thin tube of space-time that can connect two nearly flat regions far apart. It is somewhat like being at the base of a high ridge of mountains. To get to the other side, you would normally have to climb a long distance up and then back down—but not if there was a giant wormhole that cut horizontally through the rock. You could imagine creating or finding a wormhole that would lead from the vicinity of our solar system to Proxima Centauri. The distance through the wormhole might be only a few million miles, even though the earth and Proxima Centauri are twenty antiparticle moving forward in time together, there is just a single object moving in a "loop" from A to B and back again. When the object is moving forward in time (from A to B), it is called a particle. But when the object is travelling back in time (from B to A), it appears as an antiparticle travelling forward in time.

Such time travel can produce observable effects. For instance, suppose that one member of the particle/antiparticle pair (say, the antiparticle) falls into a black hole, leaving the other member without a partner with which to annihilate. The forsaken particle might fall into the hole as well, but it might also escape from the vicinity of the black hole. If so, to an observer at a distance it would appear to be a particle emitted by the black hole. You can, however, have a different but equivalent intuitive picture of the mechanism for emission of radiation from black holes. You can regard the member of the pair that fell into the black hole (say, the antiparticle) as a particle travelling backwards in time out of the hole. When it gets to the point at which the particle/antiparticle pair appeared together, it is scattered by the gravitational field of the black hole into a particle travelling forwards in time and escaping from the black hole. Or if instead it was the particle member of the pair that fell into the hole, you could regard it as an antiparticle travelling back in time and coming out of the black hole. Thus the radiation by black holes shows that quantum theory allows time travel back in time on a microscopic scale.

We can therefore ask whether quantum theory allows the possibility that, once we advance in science and technology, we might eventually manage to build a time machine. At first sight, it seems it should be possible. The Feynman sum over histories proposal is supposed to be over all histories. Thus it should include histories in which space-time is so warped that it is possible to travel into the past. Yet even if the known laws of physics do not seem to rule out time travel, there are other reasons to question whether it is possible.

One question is this: if it's possible to travel into the past, why hasn't anyone come back from the future and told us how to do it? There

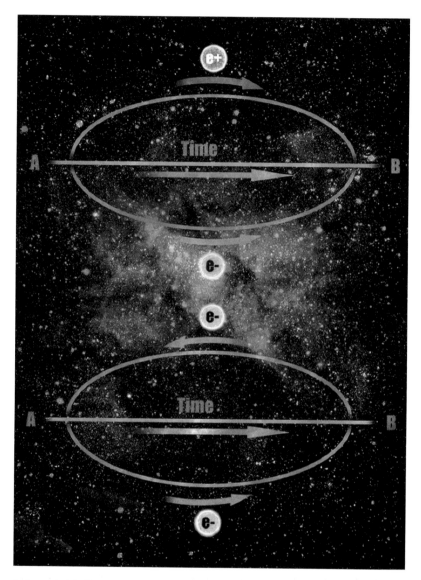

Antiparticle à la Feynman

An antiparticle can be regarded as a particle travelling backwards in time.

A virtual particle/antiparticle pair can therefore be thought of as a particle moving on a closed loop in space-time.

might be good reasons why it would be unwise to give us the secret of time travel at our present primitive state of development, but unless human nature changes radically, it is difficult to believe that some visitor from the future wouldn't spill the beans. Of course, some people would claim that sightings of UFOs are evidence that we are being visited either by aliens or by people from the future. (Given the great distance of other stars, if the aliens were to get here in reasonable time, they would need faster-than-light travel, so the two possibilities may be equivalent.) A possible way to explain the absence of visitors from the future would be to say that the past is fixed because we have observed it and seen that it does not have the kind of warping needed to allow travel back from the future. On the other hand, the future is unknown and open, so it might well have the curvature required. This would mean that any time travel would be confined to the future. There would be no chance of Captain Kirk and the starship *Enterprise* turning up at the present time.

This might explain why we have not yet been overrun by tourists from the future, but it would not avoid another type of problem, which arises if it is possible to go back and change history: why aren't we in trouble with history? Suppose, for example, someone had gone back and given the Nazis the secret of the atom bomb, or that you went back and killed your great-great-grandfather before he had children. There are many versions of this paradox, but they are essentially equivalent: we would get contradictions if we were free to change the past.

There seem to be two possible resolutions to the paradoxes posed by time travel. The first may be called the consistent histories approach. It says that even if space-time is warped so that it would be possible to travel into the past, what happens in space-time must be a consistent solution of the laws of physics. In other words, according to this viewpoint, you could not go back in time unless history already showed that you had gone back and, while there, had not killed your great-great-grandfather or committed any other acts that would conflict with the history of how you got to your current situation in the present. Moreover, when you did go back, you wouldn't be able to change recorded history; you would

merely be following it. In this view the past and future are preordained: you would not have free will to do what you wanted.

Of course, you could say that free will is an illusion anyway. If there really is a complete theory of physics that governs everything, it presumably also determines your actions. But it does so in a way that is impossible to calculate for an organism that is as complicated as a human being, and it involves a certain randomness due to quantum mechanical effects. So one way to look at it is that we say humans have free will because we can't predict what they will do. However, if a human then goes off in a rocket ship and comes back before he set off, we will be able to predict what he will do because it will be part of recorded history. Thus, in that situation, the time traveller would not in any sense have free will.

The other possible way to resolve the paradoxes of time travel might be called the alternative histories hypothesis. The idea here is that when time travellers go back to the past, they enter alternative histories that differ from recorded history. Thus they can act freely, without the constraint of consistency with their previous history. Steven Spielberg had fun with this notion in the *Back to the Future* films: Marty McFly was able to go back and change his parents' courtship to a more satisfactory history.

The alternative histories hypothesis sounds rather like Richard Feynman's way of expressing quantum theory as a sum over histories, as described in Chapter 9. This said that the universe didn't just have a single history; rather, it had every possible history, each with its own probability. However, there seems to be an important difference between Feynman's proposal and alternative histories. In Feynman's sum, each history comprises a complete space-time and everything in it. The space-time may be so warped that it is possible to travel in a rocket into the past. But the rocket would remain in the same space-time and therefore the same history, which would have to be consistent. Thus, Feynman's sum over histories proposal seems to support the consistent histories hypothesis rather than the idea of alternative histories.

We can avoid these problems if we adopt what we might call the chronology protection conjecture. This says that the laws of physics conspire to prevent macroscopic bodies from carrying information into the past. This conjecture has not been proved, but there is reason to believe it is true. The reason is that when space-time is warped enough to make time travel into the past possible, calculations employing quantum theory show that particle/antiparticle pairs moving round and round on closed loops can create energy densities large enough to give space-time a positive curvature, counteracting the warpage that allows the time travel. Because it is not yet clear whether this is so, the possibility of time travel remains open. But don't bet on it. Your opponent might have the unfair advantage of knowing the future.

· 11 ·

THE FORCES OF NATURE
AND THE UNIFICATION
OF PHYSICS

AS WAS EXPLAINED IN CHAPTER 3, it would be very difficult to construct a complete unified theory of everything in the universe all at one go. So instead we have made progress by finding partial theories that describe a limited range of happenings and by neglecting other effects or approximating them by certain numbers. The laws of science, as we know them at present, contain many numbers—for example, the size of the electric charge of the electron and the ratio of the masses of the proton and the electron—that we cannot, at the moment at least, predict from theory. Instead, we have to find them by observation and then insert them into the equations. Some call these numbers fundamental constants; others call them fudge factors.

   Whatever your point of view, the remarkable fact is that the values of these numbers seem to have been very finely adjusted to make possible the development of life. For example, if the electric charge of the electron had been only slightly different, it would have spoiled the balance of the electromagnetic and gravitational force in stars, and either they would have been unable to burn hydrogen and helium or else they would not have exploded. Either way, life could not exist. Ultimately, we would hope to find a complete, consistent, unified theory that would include all these partial theories as approximations and that did not need to be adjusted to fit the facts by picking the values of arbitrary numbers in the theory, such as the strength of the electron's charge.

The quest for such a theory is known as the unification of physics. Einstein spent most of his later years unsuccessfully searching for a unified theory, but the time was not ripe: there were partial theories for gravity and the electromagnetic force, but very little was known about the nuclear forces. Moreover, as was mentioned in Chapter 9, Einstein refused to believe in the reality of quantum mechanics. Yet it seems that the uncertainty principle is a fundamental feature of the universe we live in. A successful unified theory must, therefore, necessarily incorporate this principle.

The prospects for finding such a theory seem to be much better now because we know so much more about the universe. But we must beware of overconfidence—we have had false dawns before! At the beginning of the twentieth century, for example, it was thought that everything could be explained in terms of the properties of continuous matter, such as elasticity and heat conduction. The discovery of atomic structure and the uncertainty principle put an emphatic end to that. Then again, in 1928 physicist and Nobel Prize winner Max Born told a group of visitors to Göttingen University, "Physics, as we know it, will be over in six months." His confidence was based on the recent discovery by Dirac of the equation that governed the electron. It was thought that a similar equation would govern the proton, which was the only other particle known at the time, and that would be the end of theoretical physics. However, the discovery of the neutron and of nuclear forces knocked that one on the head too. Having said this, there are nevertheless grounds for cautious optimism that we may now be near the end of the search for the ultimate laws of nature.

In quantum mechanics, the forces or interactions between matter particles are all supposed to be carried by particles. What happens is that a matter particle, such as an electron or a quark, emits a force-carrying particle. The recoil from this emission changes the velocity of the matter particle, for the same reason that a cannon rolls back after firing a cannonball. The force-carrying particle then collides with another matter particle and is absorbed, changing the motion of that particle. The net result of the process of emission and absorption is the same as if there had been a force between the two matter particles.

Each force is transmitted by its own distinctive type of force-carrying particle. If the force-carrying particles have a high mass, it will be difficult to produce and exchange them over a large distance, so the forces they carry will have only a short range. On the other hand, if the force-carrying particles have no mass of their own, the forces will be long-range. The force-carrying particles exchanged between matter particles are said to be virtual particles because, unlike

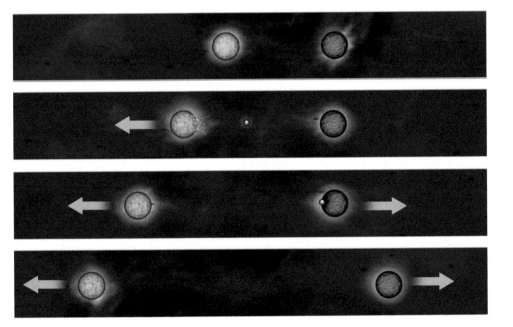

Particle Exchange
According to quantum theory, forces arise from the exchange of force-carrying particles.

real particles, they cannot be directly detected by a particle detector. We know they exist, however, because they do have a measurable effect: they give rise to forces between matter particles.

Force-carrying particles can be grouped into four categories. It should be emphasized that this division into four classes is man-made; it is convenient for the construction of partial theories, but it may not correspond to anything deeper. Ultimately, most physicists hope to find a unified theory that will explain all four forces as different aspects of a single force. Indeed, many would say this is the prime goal of physics today.

The first category is the gravitational force. This force is universal; that is, every particle feels the force of gravity, according to its mass or energy. Gravitational attraction is pictured as being caused by the exchange of virtual particles called gravitons. Gravity is the weakest of the four forces by a long way; it is so weak that we would not notice it at all were it not for two special properties that it has: it can act over large distances, and it is always attractive. This means that the very weak gravitational forces between the individual particles in two large bodies, such as the earth and the sun, can add up to produce a significant force. The other three forces either are short-range or are sometimes attractive and sometimes repulsive, so they tend to cancel out.

The next category is the electromagnetic force, which interacts with electrically charged particles such as electrons and quarks, but not with uncharged particles such as neutrinos. It is much stronger than the gravitational force: the electromagnetic force between two electrons is about a million million million million million million million (1 with forty-two zeros after it) times bigger than the gravitational force. However, there are two kinds of electric charge: positive and negative. The force between two positive charges is repulsive, as is the force between two negative charges, but the force is attractive between a positive and a negative charge.

A large body, such as the earth or the sun, contains nearly equal numbers of positive and negative charges. Thus, the attractive and repulsive forces between the individual particles nearly cancel each other out, and there is very little net electromagnetic force. However, on the small scales of atoms and molecules, electromagnetic forces dominate. The electromagnetic attraction between negatively charged electrons and positively charged protons in the nucleus causes the electrons to orbit the nucleus of the atom, just as gravitational attraction causes the earth to orbit the sun. The electromagnetic attraction is pictured as being caused by the exchange of large numbers of virtual particles called photons. Again, the photons that are exchanged are virtual particles. However, when an electron changes from one orbit to another one nearer to the nucleus, energy is released and a real photon is emitted—which can be observed as visible light by the human eye, if it has the right wavelength, or by a photon detector such as photographic film. Equally, if a real photon collides with an atom, it may move an electron from an orbit nearer the nucleus to one farther away. This uses up the energy of the photon, so it is absorbed.

The third category is called the weak nuclear force. We do not come in direct contact with this force in everyday life. It is, however, responsible for radioactivity—the decay of atomic nuclei. The weak nuclear force was not well understood until 1967, when Abdus Salam at Imperial College, London, and Steven Weinberg at Harvard both proposed theories that unified this interaction with the electromagnetic force, just as Maxwell had unified electricity and magnetism about a hundred years earlier. The predictions of the theory agreed so well with experiment that in 1979, Salam and Weinberg were awarded the Nobel Prize for physics, together with Sheldon Glashow, also at Harvard, who had suggested similar unified theories of the electromagnetic and weak nuclear forces.

The fourth category is the strongest of the four forces, the strong nuclear force. This is another force with which we don't have direct contact, but it is the force that holds most of our everyday world together. It is responsible for binding the quarks together inside the proton and neutron and for holding the protons and neutrons together in the nucleus of an atom. Without the strong force, the electric repulsion between the positively charged protons would blow apart every atomic nucleus in the universe except those of hydrogen gas, whose nuclei consist of single protons. It is believed that this force is carried by a particle, called the gluon, which interacts only with itself and with the quarks.

The success of the unification of the electromagnetic and weak nuclear forces led to a number of attempts to combine these two forces with the strong nuclear force into what is called a grand unified theory (or GUT). This title is rather an exaggeration: the resultant theories are not all that grand, nor are they fully unified, as they do not include gravity. They are also not really complete theories, because they contain a number of parameters whose values cannot be predicted from the theory but have to be chosen to fit in with experiment. Nevertheless, they may be a step towards a complete, fully unified theory.

The main difficulty in finding a theory that unifies gravity with the other forces is that the theory of gravity—general relativity—is the only one that is not a quantum theory: it does not take into account the uncertainty principle. Yet because the partial theories of the other forces depend on quantum mechanics in an essential way, unifying gravity with the other theories would require finding a way to incorporate that principle into general relativity. But no one has yet been able to come up with a quantum theory of gravity.

The reason a quantum theory of gravity has proven so hard to create has to do with the fact that the uncertainty principle means that even "empty" space is filled with pairs of virtual particles and antiparticles. If it weren't—if "empty" space were really completely empty— that would mean that all the fields, such as the gravitational and electromagnetic fields, would have to be exactly zero. However, the value of a field and its rate of change with time are like the position and velocity (i.e., change of position) of a particle: the uncertainty principle implies that the more accurately one knows one of these quantities, the less accurately one can know the other. So if a field in empty space were fixed at exactly zero, then it would have both a precise value

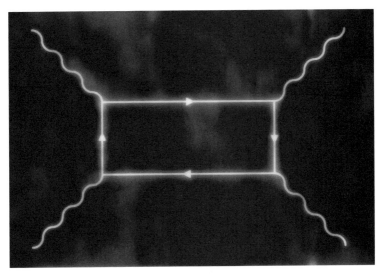

Feynman Diagram of Virtual Particle/Antiparticle Pair
The uncertainty principle, as applied to the electron, dictates that even in empty space virtual particle/antiparticle pairs appear and then annihilate each other.

(zero) and a precise rate of change (also zero), in violation of that principle. Thus there must be a certain minimum amount of uncertainty, or quantum fluctuations, in the value of the field.

One can think of these fluctuations as pairs of particles that appear together at some time, move apart, and then come together again and annihilate each other. They are virtual particles, like the particles that carry the forces: unlike real particles, they cannot be observed directly with a particle detector. However, their indirect effects, such as small changes in the energy of electron orbits, can be measured, and these data agree with the theoretical predictions to a remarkable degree of accuracy. In the case of fluctuations of the electromagnetic field, these particles are virtual photons, and in the case of fluctuations of the gravitational field, they are virtual gravitons. In the case of fluctuations of the weak and strong force fields, however, the virtual pairs are pairs of matter particles, such as electrons or quarks, and their antiparticles.

The problem is that the virtual particles have energy. In fact, because there are an infinite number of virtual pairs, they would have an infinite amount of energy and, therefore, by Einstein's equation $E=mc^2$ (see Chapter 5) they would have an infinite amount of mass. According to general relativity, this means that their gravity would curve the universe to an infinitely small size. That obviously does not happen! Similar seemingly absurd infinities occur in the other partial theories—those of the strong, weak, and electromagnetic forces— but in all these cases a process called renormalization can remove the infinities, which is why we have been able to create quantum theories of those forces.

Renormalization involves introducing new infinities that have the effect of cancelling the infinities that arise in the theory. However, they need not cancel exactly. We can choose the new infinities so as to leave small remainders. These small remainders are called the renormalized quantities in the theory.

Although in practice this technique is rather dubious mathematically, it does seem to work, and it has been used with the theories of the strong, weak, and electromagnetic forces to make predictions that agree with observations to an extraordinary degree of accuracy. Renormalization has a serious drawback from the point of view of trying to find a complete theory, though, because it means that the actual values of the masses and the strengths of the forces cannot be predicted from the theory but have to be chosen to fit the observations. Unfortunately, in attempting to use renormalization to remove the quantum infinities from general relativity, we have only two quantities that can be adjusted: the strength of gravity and the value of the cosmological constant, the term Einstein introduced into his equations because he believed that the universe was not expanding (see Chapter 7). As it turns out, adjusting these is not sufficient to remove all the infinities. We are therefore left with a quantum theory of gravity that seems to predict that certain quantities, such as the curvature of space-time, are really infinite—yet these quantities can be observed and measured to be perfectly finite!

That this would be a problem in combining general relativity and the uncertainty principle had been suspected for some time but was finally confirmed by detailed calculations in 1972. Four years later, a possible solution, called supergravity, was suggested. Unfortunately, the calculations required to find out whether or not there were any infinities left uncancelled in supergravity were so long and difficult that no one was prepared to undertake them. Even with a computer, it was reckoned, it would take many years, and the chances were very high that there would be at least one mistake, probably more. Thus we would know we had the right answer only if someone else repeated the calculation and got the same answer, and that did not seem very likely! Still, despite these problems, and the fact that the particles in the supergravity theories did not seem to match the observed particles, most scientists believed that the theory could be altered and was probably the right answer to the problem of unifying gravity with the other forces. Then in 1984 there was a remarkable change of opinion in favour of what are called string theories.

Before string theory, each of the fundamental particles was thought to occupy a single point of space. In string theories, the basic objects are not point particles but things that have a length but no other dimension, like an infinitely thin piece of string. These strings may have ends (so-called open strings) or they may be joined up with themselves in closed loops (closed strings). A particle occupies one point of space at each moment of time. A string, on the other hand, occupies a line in space at each moment of time. Two pieces of string can join together to form a single string; in the case of open strings they simply join at the ends, while in the case of closed strings it is like the two legs joining on a pair of trousers. Similarly, a single piece of string can divide into two strings.

If the fundamental objects in the universe are strings, what are the point particles we seem to observe in our experiments? In string theories, what were previously thought of as different point particles are now pictured as various waves on the string, like waves on a vibrating kite string. Yet the strings, and the vibrations along it, are so tiny that even our best technology cannot resolve their shape, so they behave, in all of our experiments, as tiny, featureless points. Imagine looking at a speck of dust: up close, or under a magnifying glass, you may find that the fleck has an irregular or even stringlike shape, yet from a distance it looks like a featureless dot.

In string theory the emission or absorption of one particle by another corresponds to the dividing or joining together of strings. For example, the gravitational force of the sun on the earth was pictured in particle theories as being caused by the emission of the force-carrying particles called gravitons by a matter particle in the sun and their absorption by a matter particle in the earth. In string theory, this process corresponds to an H-shaped tube or pipe (string theory is rather like plumbing, in a way). The two vertical sides of the H correspond to the particles in the sun and the earth, and the horizontal crossbar corresponds to the graviton that travels between them.

String theory has a curious history. It was originally invented in the late 1960s in an attempt to find a theory to describe the strong force. The idea was that particles such as the proton and the neutron could be regarded as waves on a string. The strong forces between the particles would correspond to pieces of string that went between other bits of string, as in a spider's web. For this theory to give the observed value of the strong force between particles, the strings had to be like rubber bands with a pull of about ten tons.

In 1974, Joel Scherk from the École Normale Supérieure in Paris and John Schwarz from the California Institute of Technology published a paper in which they showed that string theory could describe the nature of the gravitational force, but only if the tension in the string was about a thousand million million million million million

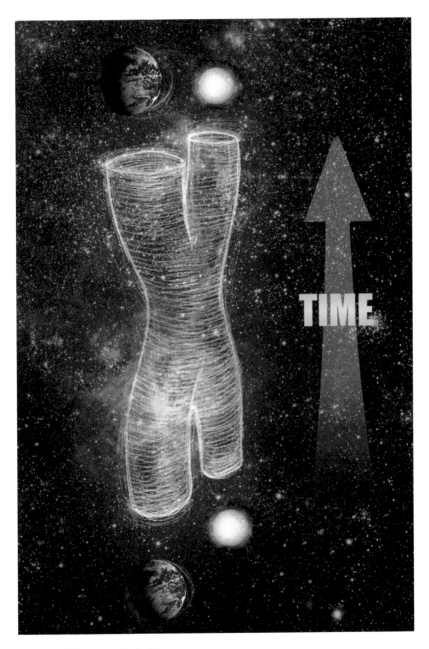

Feynman Diagrams in String Theory
In string theories, long-range forces are viewed as being caused by connecting tubes
rather than the interchange of force-carrying particles.

million tons (1 with thirty-nine zeros after it). The predictions of string theory would be just the same as those of general relativity on normal-length scales, but they would differ at very small distances, less than a thousand million million million million millionth of a centimetre (a centimetre divided by 1 with thirty-three zeros after it). Their work did not receive much attention, however, because at just about that time most people abandoned the original string theory of the strong force in favour of the theory based on quarks and gluons, which seemed to fit much better with observations. Scherk died in tragic circumstances (he suffered from diabetes and went into a coma when no one was around to give him an injection of insulin), so Schwarz was left alone as almost the only supporter of string theory, but now with the much higher proposed value of the string tension.

In 1984, interest in strings suddenly revived, apparently for two reasons. One was that people were not really making much progress towards showing that supergravity was finite or that it could explain the kinds of particles that we observe. The other was the publication of another paper by John Schwarz, this time with Mike Green of Queen Mary College, London. This paper showed that string theory might be able to explain the existence of particles that have a built-in left-handedness, like some of the particles that we observe. (The behaviour of most particles would be the same if you changed the experimental setup by reflecting it all in a mirror, but the behaviour of these particles would change. It is as if they are left- or right-handed, instead of being ambidextrous.) Whatever the reasons, a large number of people soon began to work on string theory, and a new version was developed that seemed as if it might be able to explain the types of particles that we observe.

String theories also lead to infinities, but it is thought that in the right version they will all cancel out (though this is not yet known for certain). String theories, however, have a bigger problem: they seem to be consistent only if space-time has either ten or twenty-six dimensions, instead of the usual four! Of course, extra space-time dimensions are a commonplace of science fiction. Indeed, they provide an ideal way of overcoming the normal restriction of general relativity that one cannot travel faster than light or back in time (see Chapter 10). The idea is to take a shortcut through the extra dimensions. You can picture this in the following way. Imagine that the space we live in has only two dimensions and is curved like the surface of an anchor ring or doughnut. If you were on the inside edge of the ring and you wanted to get to a point across the ring on the other side, you would have to move in a circle along the inner edge of the ring until you reached the target point. However, if you were able to travel in the third dimension, you could leave the ring and cut straight across.

Why don't we notice all these extra dimensions if they are really there? Why do we see only three space dimensions and one time dimension? The suggestion is that the other dimensions are not like the dimensions we are used to. They are curved up into a space of very small size, something like a million million million million millionth of an inch. This is so small that we just don't notice it: we see only one time dimension and three space dimensions, in which space-time is fairly flat. To picture how this works, think of the surface of a straw. If you look at it closely, you see the surface is two-dimensional. That is, the position of a point on the straw is described by two numbers, the length along the straw and the distance around the circular dimension. But its circular dimension is much smaller than its dimension of length. Because of that, if you look at the straw from a distance, you don't see the thickness of the straw and it looks one-dimensional. That is, it appears that to specify the position of a point you need only to give the length along the straw. So it is with space-time, string theorists say: on a very small scale it is ten-dimensional and highly curved, but on bigger scales you don't see the curvature or the extra dimensions.

If this picture is correct, it spells bad news for would-be space travellers: the extra dimensions would be far too small to allow a spaceship through. However, it raises a major problem for scientists as well: why should some, but not all, of the dimensions be curled up into a small ball? Presumably, in the very early universe all the dimensions would have been very curved. Why did one time dimension and three space dimensions flatten out, while the other dimensions remain tightly curled up?

One possible answer is what is called the anthropic principle, which can be paraphrased as "We see the universe the way it is because we exist." There are two versions of the anthropic principle, the weak and the strong. The weak anthropic principle states that in a universe that is large or infinite in space and/or time, the conditions necessary for the development of intelligent life will be met only in certain regions that are limited in space and time. The intelligent beings in these regions should therefore not be surprised if they observe that their locality in the universe satisfies the conditions that are necessary for their existence. It is a bit like a rich person living in a wealthy neighbourhood not seeing any poverty.

Some go much further and propose a strong version of the principle. According to this theory, there are either many different universes or many different regions of a single universe, each with its own initial configuration and, perhaps, with its own set of laws of science. In most of these universes the conditions would not be right for the development of complicated organisms; only in the few universes that are like ours would intelligent beings develop and ask the question, "Why is the universe the way we see it?" The answer is then simple: if it had been different, we would not be here!

Few people would quarrel with the validity or utility of the weak anthropic principle, but there are a number of objections that one can raise to the strong anthropic principle as an explanation of the observed state of the universe. For instance, in what sense can all these different universes be said to exist? If they are really separate from each other, what happens in another universe can have no observable consequences in our own universe. We should therefore use the principle of economy and cut them out of the theory. If, on the other hand, they were just different regions of a single universe, the laws of science would have to be the same in each region, because otherwise we could not move continuously from one region to another. In this case the only difference between the regions would be their initial configurations, so the strong anthropic principle would reduce to the weak one.

The anthropic principle gives one possible answer to the question of why the extra dimensions of string theory curled up. Two space dimensions do not seem to be enough to allow for the development of complicated beings like us. For example, two-dimensional animals living on a circle (the surface of a two-dimensional earth) would have to climb over each other in order to get past each other. And if a two-dimensional creature ate something it could not digest completely, it would have to bring up the remains the same way it swallowed them, because if there were a passage right through its body, it would divide the creature into two separate halves: our two-dimensional being would fall apart. Similarly, it is difficult to see how there could be any circulation of the blood in a two-dimensional creature.

There would also be problems with more than three space dimensions. The gravitational force between two bodies would decrease more rapidly with distance than it does in three dimensions. (In three dimensions, the gravitational force drops to one-quarter as you double the distance. In four dimensions it would drop to one-eighth, in five dimensions to one-sixteenth, and so on.) The significance of this is that the orbits of planets, like the earth, around the sun would be unstable: the least disturbance from a circular orbit (such as would be caused by the gravitational attraction of other planets) would result in the earth spiralling away from or into the sun. We would either freeze or be burned up. In fact, the same behaviour of gravity with distance in more than three space dimensions means that the sun would not be able to exist in a stable state, with pressure balancing gravity. The sun would either fall apart or collapse to form a black hole. In either case, it would not be of much use as a source of heat and light for life on earth. On a smaller scale, the electrical forces that cause the electrons to orbit around the nucleus in an atom would behave in the same way as gravitational forces. Thus the electrons would either escape from the atom altogether or would spiral into the nucleus. In either case, there would be no atoms as we know them.

It seems clear then that life, at least as we know it, can exist only in regions of space-time in which one time dimension and exactly three space dimensions are not curled up small. This would mean that we could appeal to the weak anthropic principle, provided we could show that string theory does at least allow there to be such regions of the universe—and it seems that indeed string theory does. There may well be other regions of the universe, or other universes (whatever that may mean), in which all the dimensions are curled up small or in which more than four dimensions are nearly flat, but there would be no intelligent beings in such regions to observe the different number of effective dimensions.

In addition to the question of dimensions, another problem with string theory is that there are at least five different theories (two open-string and three different closed-string theories) and millions of ways in which the extra dimensions predicted by string theory could be curled up. Why should just one string theory and one kind of curling-up be picked out? For a time there seemed no answer, and progress got bogged down. Then, starting in about 1994, people started discovering what are called dualities: different string theories and different ways of curling up the extra dimensions could lead to the same results in four dimensions. Moreover, as well as particles, which occupy a single point of space, and strings, which are lines, there were found to be other objects called p-branes, which occupied two-dimensional or higher-dimensional volumes in space. (A particle can be regarded as a 0-brane and a string as a 1-brane, but there were also p-branes for $p = 2$ to $p = 9$. A 2-brane can be thought of as something like a two-dimensional membrane. It is harder to picture the higher-dimensional branes.) What this seems to indicate is that there is a sort of democracy (in the sense of having equal voices) among supergravity, string, and p-brane theories: they seem to fit together, but none can be said to be more fundamental than the others. Instead they all appear to be different approximations to some more fundamental theory, each valid in different situations.

People have searched for this underlying theory, but without any success so far. It is possible that there may not be any single formulation of the fundamental theory any more than, as Gödel showed, one could formulate arithmetic in terms of a single set of axioms. Instead it may be like maps—you can't use a single flat map to describe the round surface of the earth or the surface of an anchor ring: you need at least two maps in the case of the earth and four for the anchor ring to cover

The Importance of Being Three-Dimensional
In more than three space dimensions, planetary orbits would be unstable and planets would either fall into the sun or escape its attraction altogether.

always a degree of uncertainty. If you like, you could ascribe this randomness to the intervention of God. But it would be a very strange kind of intervention, with no evidence that it is directed towards any purpose. Indeed, if it were, it would by definition not be random. In modern times, we have effectively removed the third possibility above by redefining the goal of science: our aim is to formulate a set of laws that enables us to predict events only up to the limit set by the uncertainty principle.

The second possibility, that there is an infinite sequence of more and more refined theories, is in agreement with all our experience so far. On many occasions we have increased the sensitivity of our measurements or made a new class of observations, only to discover new phenomena that were not predicted by the existing theory, and to account for these we have had to develop a more advanced theory. By studying particles that interact with more and more energy, we might indeed expect to find new layers of structure more basic than the quarks and electrons that we now regard as "elementary" particles.

Gravity may provide a limit to this sequence of "boxes within boxes." If we had a particle with an energy above what is called the Planck energy, its mass would be so concentrated that it would cut itself off from the rest of the universe and form a little black hole. Thus it does seem that the sequence of more and more refined theories should have some limit as we study higher and higher energies, so there should be some ultimate theory of the universe. Yet the Planck energy is a very long way from the energies we can produce in the laboratory at the present time. We shall not bridge that gap with particle accelerators in the foreseeable future. The very early stages of the universe, however, are an arena where such energies must have occurred. There is a good chance that the study of the early universe and the requirements of mathematical consistency will lead us to a complete unified theory within the lifetime of some of us who are around today, always presuming we don't blow ourselves up first!

What would it mean if we actually did discover the ultimate theory of the universe?

As was explained in Chapter 3, we could never be quite sure that we had indeed found the correct theory, since theories can't be proved. But if the theory was mathematically consistent and always gave predictions that agreed with observations, we could be reasonably confident that it was the right one. It would bring to an end a long and glorious chapter in the history of humanity's intellectual struggle to understand the universe. But it would also revolutionize the ordinary person's understanding of the laws that govern the universe.

In Newton's time, it was possible for an educated person to have a grasp of the whole of human knowledge, at least in broad strokes. But since then, the pace of the development of science has made this impossible. Because theories are always being changed to account for new observations, they are never properly digested or simplified so that ordinary people can understand them. You have to be a specialist, and even then you can only hope to have a proper grasp of a small proportion of the scientific theories. Further, the rate of progress is so rapid that what you learn at school or university is always a bit out of date. Only a few people can keep up with the rapidly advancing frontier of knowledge, and they have to devote their whole time to it and specialize in a small area. The rest of the population has little idea of the advances that are being made or the excitement they are generating. On the other hand, seventy years ago, if Eddington is to be believed, only two people understood the general theory of relativity. Nowadays tens of thousands of university graduates do, and many millions of people are at least familiar with the idea. If a complete unified theory is discovered, it will be only a matter of time before it becomes digested and simplified in the same way and taught in schools, at least in outline. We will then all be able to have some understanding of the laws that govern the universe and are responsible for our existence.

Even if we do discover a complete unified theory, though, it would not mean that we would be able to predict events in general, for two reasons. The first is the limitation that the uncertainty principle of quantum mechanics sets on our powers of prediction. There is nothing we can do to get around that. In practice, however, this first limitation is less restrictive than the second one. It arises from the fact that we most likely could not solve the equations of such a theory, except in very simple situations. As we've said, no one can solve exactly the quantum equations for an atom consisting of a nucleus plus more than one electron. We can't even solve exactly the motion of three bodies in a theory as simple as Newton's theory of gravity, and the difficulty increases with the number of bodies and the complexity of the theory. Approximate solutions usually suffice for applications, but they hardly meet the grand expectations aroused by the term "unified theory of everything"!

Today, we already know the laws that govern the behaviour of matter under all but the most extreme conditions. In particular, we know the basic laws that underlie all of chemistry and biology. Yet we have certainly not reduced these subjects to the status of solved problems. And we have had, as yet, little success in predicting human behaviour from mathematical equations! So even if we do find a complete set of basic laws, there will still be in the years ahead the intellectually challenging task of developing better approximation methods so that we can make useful predictions of the probable outcomes in complicated and realistic situations. A complete, consistent, unified theory is only the first step: our goal is a complete understanding of the events around us, and of our own existence.

· 12 ·

CONCLUSION

WE FIND OURSELVES IN A BEWILDERING world. We want to make sense of what we see around us and to ask: What is the nature of the universe? What is our place in it, and where did it and we come from? Why is it the way it is?

To try to answer these questions, we adopt some picture of the world. Just as an infinite tower of tortoises supporting the flat earth is such a picture, so is the theory of superstrings. Both are theories of the universe, though the latter is much more mathematical and precise than the former. Both theories lack observational evidence: no one has ever seen a giant tortoise with the earth on its back, but then, no one has ever seen a superstring either. However, the tortoise theory fails to be a good scientific theory because it predicts that people should be able to fall off the edge of the world. This has not been found to agree with experience, unless that turns out to be the explanation for the people who are supposed to have disappeared in the Bermuda Triangle!

The earliest theoretical attempts to describe and explain the universe involved the idea that events and natural phenomena were controlled by spirits with human emotions who acted in a very

From Turtles to Curved Space
Ancient and modern views of the universe.

humanlike and unpredictable manner. These spirits inhabited natural objects, such as rivers, mountains, and celestial bodies including the sun and moon. They had to be placated and their favour sought in order to ensure the fertility of the soil and the rotation of the seasons. Gradually, however, it must have been noticed that there were certain regularities: the sun always rose in the east and set in the west, whether or not a sacrifice had been made to the sun god. Further, the sun, the moon, and the planets followed precise paths across the sky that could be predicted in advance with considerable accuracy. The sun and the moon might still be gods, but they were gods who obeyed strict laws, apparently without any exceptions, if one discounts stories such as that of the sun stopping for Joshua.

At first, these regularities and laws were obvious only in astronomy and a few other situations. However, as civilization developed, and particularly in the last three hundred years, more and more regularities and laws were discovered. The success of these laws led Laplace at the beginning of the nineteenth century to postulate scientific determinism; that is, he suggested that there would be a set of laws that would determine the evolution of the universe precisely, given its configuration at any one time.

Laplace's determinism was incomplete in two ways: it did not say how the laws should be chosen, and it did not specify the initial configuration of the universe. These were left to God. God would choose how the universe began and what laws it obeyed, but He would not intervene in the universe once it had started. In effect, God was confined to the areas that nineteenth-century science did not understand.

We now know that Laplace's hopes of determinism cannot be realized, at least in the terms he had in mind. The uncertainty principle of quantum mechanics implies that certain pairs of quantities, such as the position and velocity of a particle, cannot both be predicted with complete accuracy. Quantum mechanics deals with this situation via a class of quantum theories in which particles don't have well-defined positions and velocities but are represented by a wave. These quantum theories are deterministic in the sense that they give laws for the evolution of the wave with time. Thus if we know the wave at one time, we can calculate it at any other time. The unpredictable, random element comes in only when we try to interpret the wave in terms of the positions and velocities of particles. But maybe that is our mistake: maybe there are no particle positions and velocities, but only waves. It is just that we try to fit the waves to our preconceived ideas of positions and velocities. The resulting mismatch is the cause of the apparent unpredictability.

In effect, we have redefined the task of science to be the discovery of laws that will enable us to predict events up to the limits set by the uncertainty principle. The question remains, however: how or why were the laws and the initial state of the universe chosen?

This book has given special prominence to the laws that govern gravity, because it is gravity that shapes the large-scale structure of the universe, even though it is the weakest of the four categories of forces. The laws of gravity were incompatible with the view, held until quite recently, that the universe is unchanging in time: the fact that gravity is always attractive implies that the universe must be either expanding or contracting. According to the general theory of relativity, there must have been a state of infinite density in the past, the big bang, which would have been an effective beginning of time. Similarly, if the whole universe collapsed, there must be another state of infinite density in the future, the big crunch, which would be an end of time. Even if the whole universe did not collapse, there would be singularities in any localized regions that collapsed to form black holes. These singularities would be an end of time for anyone who fell into the black hole. At the big bang and other singularities, all the laws would have broken down, so God would still have had complete freedom to choose what happened and how the universe began.

When we combine quantum mechanics with general relativity, there seems to be a new possibility that did not arise before: that space and time together might form a finite, four-dimensional space without singularities or boundaries, like the surface of the earth but with more dimensions. It seems that this idea could explain many of the observed features of the universe, such as its large-scale uniformity and also the smaller-scale departures from homogeneity, including galaxies, stars, and even human beings. But if the universe is completely self-contained, with no singularities or boundaries, and completely described by a unified theory, that has profound implications for the role of God as creator.

Einstein once asked, "How much choice did God have in constructing the universe?" If the no-boundary proposal is correct, God had no freedom at all to choose initial conditions. God would, of course, still have had the freedom to choose the laws that the universe obeyed. This, however, may not really have been all that much of a choice; there may well be only one, or a small number, of complete unified theories, such as string theory, that are self-consistent and allow the existence of structures as complicated as human beings who can investigate the laws of the universe and ask about the nature of God.

Even if there is only one possible unified theory, it is just a set of rules and equations. What is it that breathes fire into the equations and makes a universe for them to describe? The usual approach of science of constructing a mathematical model cannot answer the questions of why there should be a universe for the model to describe. Why does the universe go to all the bother of existing? Is the unified theory so compelling that it brings about its own existence? Or does it need a creator, and if so, does He have any other effect on the universe? And who created Him?

Up to now, most scientists have been too occupied with the development of new theories that describe what the universe is to ask why. On the other hand, the people whose business it is to ask why, the philosophers, have not been able to keep up with the advance of scientific theories. In the eighteenth century, philosophers considered the whole of human knowledge, including science, to be their field and discussed questions such as whether the universe had a beginning. However, in the nineteenth and twentieth centuries, science became too technical and mathematical for the philosophers, or anyone else except a few specialists. Philosophers reduced the scope of their inquiries so much that Wittgenstein, the most famous philosopher of the twentieth century, said, "The sole remaining task for philosophy is the analysis of language." What a comedown from the great tradition of philosophy from Aristotle to Kant!

If we do discover a complete theory, it should in time be understandable in broad principle by everyone, not just a few scientists. Then we shall all, philosophers, scientists, and just ordinary people, be able to take part in the discussion of the question of why it is that we and the universe exist. If we find the answer to that, it would be the ultimate triumph of human reason—for then we would know the mind of God.

Albert Einstein

EINSTEIN'S CONNECTION WITH THE POLITICS of the nuclear bomb is well known: he signed the famous letter to President Franklin Roosevelt that persuaded the United States to take the idea seriously, and he engaged in postwar efforts to prevent nuclear war. But these were not just the isolated actions of a scientist dragged into the world of politics. Einstein's life was, in fact, to use his own words, "divided between politics and equations".

Einstein's earliest political activity came during the First World War, when he was a professor in Berlin. Sickened by what he saw as the waste of human lives, he became involved in anti-war demonstrations. His advocacy of civil disobedience and public encouragement of people to refuse conscription did little to endear him to his colleagues. Then, following the war, he directed his efforts towards reconciliation and improving international relations. This too did not make him popular, and soon his politics were making it difficult for him to visit the United States, even to give lectures.

Einstein's second great cause was Zionism. Although he was Jewish by descent, Einstein rejected the biblical idea of God. However, a growing awareness of anti-Semitism, both before and during the First World War, led him gradually to identify with the Jewish community, and later to become an outspoken supporter of Zionism. Once more, unpopularity did not stop him from speaking his mind. His theories came under attack; an anti-Einstein organization was even set up. One man was convicted of inciting others to murder Einstein (and fined a mere six dollars). But Einstein was phlegmatic. When a book was published entitled *100 Authors Against Einstein,* he retorted, "If I were wrong, then one would have been enough!"

In 1933, Hitler came to power. Einstein was in America and declared he would not return to Germany. Then, while Nazi militia raided his house and confiscated his bank account, a Berlin newspaper displayed the headline "Good News from Einstein—He's Not Coming Back." In the face of the Nazi threat, Einstein renounced pacifism, and eventually, fearing that German scientists would build a nuclear bomb, he proposed that the United States should develop its own. But even before the first atomic bomb had been detonated, he was publicly warning of the dangers of nuclear war and proposing international control of nuclear weaponry.

Throughout his life, Einstein's efforts towards peace probably achieved little that would last—and certainly won him few friends. His vocal support of the Zionist cause, however, was duly recognized in 1952, when he was offered the presidency of Israel. He declined, saying he thought he was too naive about politics. But perhaps his real reason was different: to quote him again, "Equations are more important to me, because politics is for the present, but an equation is something for eternity."

• Galileo Galilei •

GALILEO, PERHAPS MORE THAN ANY OTHER single person, was responsible for the birth of modern science. His renowned conflict with the Catholic Church was central to his philosophy, for Galileo was one of the first to argue that man could hope to understand how the world works and, moreover, that we could do this by observing the real world. Galileo had believed Copernican theory (that the planets orbited the sun) since early on, but it was only when he found the evidence needed to support the idea that he started to publicly espouse it. He wrote about Copernicus's theory in Italian (not the usual academic Latin), and soon his views became widely adopted outside the universities. This annoyed the Aristotelian professors, who united against him, seeking to persuade the Catholic Church to ban Copernicanism.

Galileo, worried by this, travelled to Rome to speak to the ecclesiastical authorities. He argued that the Bible was not intended to tell us anything about scientific theories and that it was usual to assume that, where the Bible conflicted with common sense, it was being allegorical.

But the Church was afraid of a scandal that might undermine its fight against Protestantism, and so took repressive measures. It declared Copernicanism "false and erroneous" in 1616, and commanded Galileo never again to "defend or hold" the doctrine. Galileo acquiesced.

In 1623, a longtime friend of Galileo's became the pope. Immediately Galileo tried to get the 1616 decree revoked. He failed, but he did manage to get permission to write a book discussing both Aristotelian and Copernican theories, on two conditions: he would not take sides, and he would come to the conclusion that man could in any case not determine how the world worked because God could bring about the same effects in ways unimagined by man, who could not place restrictions on God's omnipotence.

The book, *Dialogue Concerning the Two Chief World Systems,* was completed and published in 1632, with the full backing of the censors—and was immediately greeted throughout Europe as a literary and philosophical masterpiece. Soon the pope, realizing that people were seeing the book as a convincing argument in favour of Copernicanism, regretted having allowed its publication. The pope argued that although the book had the official blessing of the censors, Galileo had nevertheless contravened the 1616 decree. He brought Galileo before the Inquisition, which sentenced him to house arrest for life and commanded him to publicly renounce Copernicanism. For a second time, Galileo acquiesced.

Galileo remained a faithful Catholic, but his belief in the independence of science had not been crushed. Four years before his death in 1642, while he was still under house arrest, the manuscript of his second major book was smuggled to a publisher in Holland. It was this work, referred to as *Two New Sciences,* even more than his support for Copernicus, that was to be the genesis of modern physics.

Isaac Newton

ISAAC NEWTON WAS NOT A PLEASANT man. His relations with other academics were notorious, with most of his later life spent embroiled in heated disputes. Following publication of *Principia Mathematica*—surely the most influential book ever written in physics—Newton rose rapidly into public prominence. He was appointed president of the Royal Society and became the first scientist ever to be knighted.

Newton soon clashed with the Astronomer Royal, John Flamsteed, who had earlier provided him with much-needed data for *Principia* but was now withholding information that Newton wanted. Newton would not take no for an answer: he had himself appointed to the governing body of the Royal Observatory and then tried to force immediate publication of the data. Eventually he arranged for Flamsteed's work to be seized and prepared for publication by Flamsteed's mortal enemy, Edmond Halley. But Flamsteed took the case to court and, in the nick of time, won a court order preventing distribution of the stolen work. Newton was incensed and sought his revenge by systematically deleting all references to Flamsteed in later editions of *Principia*.

A more serious dispute arose with the German philosopher

Gottfried Leibniz. Both Leibniz and Newton had independently developed a branch of mathematics called calculus, which underlies most of modern physics. Although we now know that Newton discovered calculus years before Leibniz, he published his work much later. A major row ensued over who had been first, with scientists vigorously defending both contenders. It is remarkable, however, that most of the articles appearing in defence of Newton were originally written by his own hand, though published under the names of friends! As the row grew, Leibniz made the mistake of appealing to the Royal Society to resolve the dispute. Newton, as president, appointed an "impartial" committee to investigate, coincidentally consisting entirely of his friends! But that was not all: Newton then wrote the committee's report himself and had the Royal Society publish it, officially accusing Leibniz of plagiarism. Still unsatisfied, he then wrote an anonymous review of the report in the Royal Society's own periodical. Following the death of Leibniz, Newton is reported to have declared that he had taken great satisfaction in "breaking Leibniz's heart".

During the period of these two disputes, Newton had already left Cambridge and academe. He had been active in anti-Catholic politics at Cambridge and later in Parliament, and was rewarded eventually with the lucrative post of Warden of the Royal Mint. Here he used his talents for deviousness and vitriol in a more socially acceptable way, successfully conducting a major campaign against counterfeiting, even sending several men to their death on the gallows.

Glossary

Absolute zero: The lowest possible temperature, at which substances contain no heat energy.

Acceleration: The rate at which the speed of an object is changing.

Anthropic principle: The idea that we see the universe the way it is because if it were different, we would not be here to observe it.

Antiparticle: Each type of matter particle has a corresponding antiparticle. When a particle collides with its antiparticle, both are annihilated, leaving only energy.

Atom: The basic unit of ordinary matter, made up of a tiny nucleus (consisting of protons and neutrons) surrounded by orbiting electrons.

Big bang: The singularity at the beginning of the universe.

Big crunch: The singularity at the end of the universe.

Black hole: A region of space-time from which nothing, not even light, can escape, because gravity is so strong.

Coordinates: Numbers that specify the position of a point in space and time.

Cosmological constant: A mathematical device used by Einstein to give space-time an inbuilt tendency to expand.

Cosmology: The study of the universe as a whole.

Dark matter: Matter in galaxies, clusters, and possibly between clusters that has not been observed directly but can be detected by its gravitational effect. As much as 90 per cent of the mass of the universe may be in the form of dark matter.

Duality: A correspondence between apparently different theories that lead to the same physical results.

Einstein–Rosen bridge: A thin tube of space-time linking two black holes. *See also* **Wormhole.**

Electric charge: A property of a particle by which it may repel (or attract) other particles that have a charge of similar (or opposite) sign.

Electromagnetic force: The force that arises between particles with electric charge; the second strongest of the four fundamental forces.

Electron: A particle with negative electric charge that orbits the nucleus of an atom.

Electroweak unification energy: The energy (around 100 GeV) above which the distinction between the electromagnetic force and the weak force disappears.

Elementary particle: A particle that, it is believed, cannot be sub-divided.

Event: A point in space-time, specified by its time and place.

Event horizon: The boundary of a black hole.

Field: Something that exists throughout space and time, as opposed to a particle that exists at only one point at a time.

Frequency: For a wave, the number of complete cycles per second.

Gamma rays: Electromagnetic rays of very short wavelength, pro-duced in radioactive decay or by collisions of elementary particles.

General relativity: Einstein's theory based on the idea that the laws of science should be the same for all observers, no matter how they are moving. It explains the force of gravity in terms of the curva-ture of a four-dimensional space-time.

Geodesic: The shortest (or longest) path between two points.

Grand unified theory (GUT): A theory that unifies the electro-magnetic, strong, and weak forces.

Light-second (light-year): The distance travelled by light in one second (year).

Magnetic field: The field responsible for magnetic forces, now incor-porated along with the electric field into the electromagnetic field.

Mass: The quantity of matter in a body; its inertia, or resistance to acceleration.

Microwave background radiation: The radiation from the glow-ing of the hot early universe, now so greatly red-shifted that it appears not as light but as microwaves (radio waves with a wave-length of a few centimetres).

Neutrino: An extremely light particle that is affected only by the weak force and gravity.

Neutron: A particle very similar to the proton but without charge, which accounts for roughly half the particles in the nuclei of most atoms.

Neutron star: The cold star that sometimes remains after a super-nova explosion, when the core of material at the centre of a star collapses into a dense mass of neutrons.

No-boundary condition: The idea that the universe is finite but has no boundary.

Nuclear fusion: The process by which two nuclei collide and coa-lesce to form a single, heavier nucleus.

Nucleus: The central part of an atom, consisting only of protons and neutrons, held together by the strong force.

Particle accelerator: A machine that, using electromagnets, can accelerate moving charged particles, giving them more energy.

Phase: For a wave, the position in its cycle at a specified time: a mea-sure of whether it is at a crest, a trough, or somewhere in between.

Photon: A quantum of light.

Planck's quantum principle: The idea that light (or any other classical waves) can be emitted or absorbed only in discrete quanta, whose energy is proportional to their frequency, and inversely proportional to their wavelength.

Positron: The (positively charged) antiparticle of the electron.

Proportional: "X is proportional to Y" means that when Y is multiplied by any number, so is X. "X is inversely proportional to Y" means that when Y is multiplied by any number, X is divided by that number.

Proton: A particle very similar to the neutron but positively charged, which accounts for roughly half the particles in the nuclei of most atoms.

Quantum mechanics: The theory developed from Planck's quantum principle and Heisenberg's uncertainty principle.

Quark: A (charged) elementary particle that feels the strong force. Protons and neutrons are each composed of three quarks.

Radar: A system using pulsed radio waves to detect the position of objects by measuring the time it takes a single pulse to reach the object and be reflected back.

Radioactivity: The spontaneous breakdown of one type of atomic nucleus into another.

Red shift: The reddening of light from a star that is moving away from us, due to the Doppler effect.

Singularity: A point in space-time at which the space-time curvature (or some other physical quantity) becomes infinite.

Space-time: The four-dimensional space whose points are events.

Spatial dimension: Any of the three dimensions—that is, any dimension except the time dimension.

Special relativity: Einstein's theory based on the idea that the laws of science should be the same for all observers, no matter how they are moving, in the absence of gravitational phenomena.

Spectrum: The component frequencies that make up a wave. The visible part of the sun's spectrum can be seen in a rainbow.

String theory: A theory of physics in which particles are described as waves on strings. Strings have length but no other dimension.

Strong force: The strongest of the four fundamental forces, with the shortest range of all. It holds the quarks together within protons and neutrons, and holds the protons and neutrons together to form atoms.

Uncertainty principle: The principle, formulated by Heisenberg, that it is not possible to be exactly sure of both the position and the velocity of a particle; the more accurately one is known, the less accurately the other can be known.

Virtual particle: In quantum mechanics, a particle that can never be directly detected, but whose existence does have measurable effects.

Wave/particle duality: The concept in quantum mechanics that there is no distinction between waves and particles; particles may sometimes behave like waves, and waves like particles.

Wavelength: For a wave, the distance between two adjacent troughs or two adjacent crests.

Weak force: The second weakest of the four fundamental forces, after gravity, with a very short range. It affects all matter particles, but not force-carrying particles.

Weight: The force exerted on a body by a gravitational field. It is proportional to, but not the same as, its mass.

Wormhole: A thin tube of space-time connecting distant regions of the universe. Wormholes might also link to parallel or baby universes and could provide the possibility of time travel.

About the Authors

STEPHEN HAWKING is Lucasian Professor of Mathematics at the University of Cambridge; his other books for the general reader include the essay collection *Black Holes and Baby Universes* and *The Universe in a Nutshell*.

LEONARD MLODINOW, his collaborator for this new edition, has taught at Caltech, written for *Star Trek: The Next Generation,* and is the author of *Euclid's Window* and *Feynman's Rainbow* and the co-author of the children's book series The Kids of Einstein Elementary.

· Index ·